A Guide to Climate Change Lunacy

bad forecasting, terrible solutions

Mark Lawson

connorcourt
PUBLISHING

Published in 2010 by Connor Court Publishing Pty Ltd

Copyright © Mark Lawson

All rights reserved

Connor Court Publishing Pty Ltd
PO BOX 1,
BALLAN, VIC, 3342
www.connorcourt.com
sales@connorcourt.com

Cover design by Ian James - ianjgd@bigpond.net.au
Front cover photo from IStockphoto

Printed in Australia by Openbook/Howden

National Library of Australia Cataloguing-in-Publication entry

Lawson, Mark

A guide to climate change lunacy / Mark Lawson

ISBN: 9781921421426 (pbk.)

Notes: Includes index

Subjects: Climatic changes
Climatic changes--Government policy
Global warming
Global warming--Government policy
Wind power plants

Dewey Number: 363.73874

To my mum who passed away while this book was being written.

CONFESSIONS OF AN ECO-CRIMINAL

When I wrote a piece for the site Onlineopinion (www.onlineopinion.com.au) pointing out that the scientists in the global warming camp seemed to have come around to the view that temperatures were not behaving as expected, and were offering a range of different explanations, a part of one response is so over the top it is worth quoting.

"Further, your denialist side includes a significant proportion of the largest polluters on the planet, however, you appear happy to consort with known eco-criminals, refusing to acknowledge that anthropogenic pollution includes anthropogenic emissions of CO_2 too – that's scientific. Do you know the difference between a VOC and a sock?" (VOC stands for Volatile Organic Compound; a category of pollutant. It has no relevance to the climate debate.)

This type of abuse, so extreme that it amuses rather than offends the target, is quite common from the less disciplined climate activists, but okay, call me an eco-criminal. No doubt I will be called a lot worse. I have even set up an email address ecocriminal@optusnet.com.au for those who need to vent their feelings on a hapless sceptic who does not understand the big picture, or at least the big picture as seen by activists.

A few disclaimers. About 25 years ago and counting I was a member of the Young Liberals in Victoria, and at one point was on the state executive. I have not been a member of any political party since, although my family has some connections with the Liberal

Party. The only other organisation I have joined, apart from what is now the journalists section of the union the Media and Entertainment Arts Alliance, is the Australian Skeptics. My only conection with the energy industry, apart from occasionally reporting on it over the years, is when I pay electricity bills or fill up the family car. My employer, the *Australian Financial Review*, has no connection with this book. The views in it are my own and not those of the AFR, and most emphatically not those of most of my colleagues.

There are people who have consciously helped and I would like to thank them but I am reluctant to name them in case they also become targets for activist vitriol. Two online newsletters which have proved invaluable as sources of information are CCNet compiled by Benny Peiser and The Week That Was, founded by S. Fred Singer and now compiled by Kenneth Haapala. Both Peiser and Singer feature elsewhere in this book. Then there are those who have helped me unconsciously, particularly those bitter critics in online opinion and the Australian Skeptics, plus the occasional AFR reader, who have objected to the various articles I have written. Most of the responses have been abuse similar to the material quoted at the beginning of this note, but there have been a few gems among the dross.

For those of you who see red while reading this book remember that abuse and ranting will have no effect, but I am always interested in reasoned responses.

CONTENTS

Introducting Lunacy   5
1. Forecasting Follies   13
2. Ice Age Escapades   33
3. Bad Calls   57
4. Model Mayhem   75
5. Confusion Past and Present   99
6. Numbers That Don't Add Up   113
7. Cutting Carbon   131
8. Blowing In The Wind   147
9. Sea, Floods and Ice   167
10. Acid and Adaption   179
11. Storms and Diseases   191
12. Slippery Oil Numbers   199
13. Show Me The Money!   215
Index   223

4

INTRODUCING LUNACY

For the past million years, the earth's climate has repeatedly gone through a distinct 100,000 year cycle of a long, cold ice age punctuated by a brief warm, interglacial period such as the one we are in now. Global temperatures vary by perhaps 14 degrees centigrade from the bottom of the ice age to the warmest part of the interglacial. In the two million years before that, the earth's climate cycled between ice ages and interglacials in much the same way, but with a period just 40,000 years long and generally higher temperatures.

Scientists do not know why the earth's climate switched from a 40,000 year to a 100,000 year cycle, or why the earth has been cooling down over the past few million years as part of its present "ice age" phase. There have been "hot house" phases in the more distant past. There is some vague theory to explain the end of the ice ages that involves slow periodic changes in the earth's orbit and tilt of its poles (the same tilt that causes the seasons), plus increases in carbon dioxide levels. But both elements of this explanation have run into severe timing difficulties in recent years, with field research indicating that the events happen before the supposed causes.

The theory concerning the end of interglacials and the sudden plunge into ice ages is even vaguer and there is nothing in conventional theory to indicate why our own interglacial, known as the Holocene, has lasted 10,000 years plus. The preceding three interglacials were all much shorter. There are many more problems.

In the interglacial before this one, about 130,000 years ago, temperatures are known to have been two degrees or so above present levels, or about the middle of the Intergovernmental Panel on Climate Change's range of forecasts for the next 100 years. At the time, at least to judge from measurements taken from air trapped inside ancient ice, carbon dioxide levels were around 300 parts per million (0.03 per cent of the atmosphere) or well below the 360 ppm which activists insist is necessary to avoid damaging climate change. CO_2 concentrations were still at 300 ppm when temperatures plummeted eight degrees or so and the earth fell into the ice age before our own Holocene.

When temperatures started to increase in the run up to our own Holocene, the humans who by then had spread over most of the globe were major beneficiaries. The first towns and walls date from about 10,000 years ago, and the first of the culture that earned the term civilisation, the Sumerians, formed in what is now Iraq 5,000 years ago. But the Sumerians and succeeding civilisations were all affected by changes in climate zones, rainfall and overall temperatures the reasons for which are only dimly understood, although some pieces of the puzzle are now falling into place. Global temperatures, for example, are known to have varied on a 1,500 year cycle (give or take 500 years) – a cycle, which has been tracked back through hundreds of thousands of years and is independent of the ice age-interglacial cycle. This cycle resulted in higher temperatures in the early Holocene, higher temperatures in Roman times, followed by the colder period of the dark ages, then a Medieval Warm Period and the Little Ice Age and, finally, the modern warm period.

The Little Ice Age is thought to have only ended in 1850 and the subsequent increase in temperatures, about eight tenths of a degree in 160 years, or perhaps just six-tenths if we rely on satellite measurements of temperature, is small change in the climate history of the earth. But the increase happened about the same time as the rise and rise of industry and CO_2 concentrations in the atmosphere are known to be have increased. The Intergovernmental Panel on Climate Change formed in the late 1980s over concern about rising temperatures has issued a series of reports linking the increase in CO_2 concentrations to that of temperature. These reports draw on gigantic computer models which try to simulate the behaviour of the earth's atmosphere and ocean using what scientists know, or think they know, about the interaction of solar radiation, clouds, land, ocean currents, trade winds, ice sheets, and CO_2 concentrations, to name just a few factors. The latest versions of this immense digital effort forecast increases in temperatures of more than four degrees, with a best guess of around three degrees, over the next century.

Consumers of these forecasts, including even some scientists not directly involved with this modelling work, seem to think that these projections are the result of brave climate modellers punching a few numbers into a well-established climate model, run on a super computer, then sauntering around to the printer to examine the results. Nothing could be further from the truth. The models are extremely sensitive to starting conditions (the initial state of the atmosphere and ocean), have to be run multiple times to get an agreed projection, and some models still require the occasional "flux adjustment" (unkind people would call this fudging) to keep on track.

In addition, our brave climate modellers have to make a number of assumptions about climate. Some of these are built into the models but others are there simply because nothing is known about that aspect of climate. A crucial set of assumptions involves clouds and humidity, which even the scientists strongly defending the projections may not be aware of, just as the vast mass of green enthusiasts may not be aware that the supposed warming is an indirect result of increases in carbon dioxide, not a direct result. There is a crucial assumption concerning water vapour. But then there are many assumptions in those models and a wrong step in any of them may (or may not) completely derail the projections. That is because projections and forecasts of any kind are very hard to get right, and very easy to derail.

These computer projections of climate, in turn, depend on another set of projections of methane, CO_2 and industrial gas concentrations in the atmosphere. These were controversial even when they were released, are now a decade old and already hopelessly wrong as far as methane is concerned. Methane concentrations in the atmosphere stopped increasing around the turn of the century. No one knows why. CO_2 concentrations are near the bottom end of those projections, despite repeated assurances that industrial emissions are well ahead of projections, and all this after only a decade of real world results for projections that run over a full century. Once we get past those problems there remain niggling doubts, which the IPCC has done its best to ignore, that its view of the carbon cycle and the behaviour of carbon dioxide in the atmosphere may be entirely wrong in the first place.

Then there is the extensive reliance on computer models. Before their widespread use in climate studies, these models also had no great name in science. They can be invaluable if handled properly, but this sort of large scale use to make projections a full century out is almost unprecedented. The precedents that do exist are far from encouraging. A few scientists will also tell you that modelling natural systems is impossible; they are too complicated. One possible exception is weather forecasting but that is in part because meteorologists can now see weather patterns developing with satellites and instrument networks. Far more importantly they have gained a lot of experience by getting forecasts wrong.

In fact, any form of forecasting is fraught with difficulty, whether the forecast is about the next solar cycle, tomorrow's share market prices, crime rates in the next decade, electricity prices in five years, the total catch size fishing boats can take from a fishery without affecting the natural resource, or whether next winter will be mild or severe. If you have a forecasting system with an established track record then fine, if not it's best not to say anything, particularly for a natural system. You will be proved wrong. The same rule is known to apply to experts, irrespective of whether they are lone wolves or are hunting in packs. There is also nothing in the history of science to suggest that the number of scientists for or against a particular proposition matters in the slightest unless, as noted, they have a proven track record in a particular discipline. Then it matters, mostly. They cannot point to any forecasting track record in climate studies.

In an attempt to give climate forecasts some veracity, the IPCC has a system of what it calls "peer review" which does not touch any of the key assumptions in the computer models and is meaningless as a check on forecasting. In any case, as this book was being written, it became increasingly evident that a number of the scientists who compiled the IPCC report grabbed whatever material they could find to put in the report, and none of that was picked up by the review system.

Despite these well-known problems with forecasting anything in general, and of forecasting in non-linear chaotic system of extraordinary complexity in particular, some scientists and a lot of activists insist that there is no doubt about the IPCC forecasts. Those familiar with the history of forecasting will realise that this is akin to the captain of a cruise ship sailing full tilt through iceberg-infested waters calling the passengers together specifically to tell them that the ship is unsinkable. Activist assurances that the overwhelming majority of scientists support the IPCC position are not only quite wrong – there is no such consensus – but akin to our hypothetical passengers also being told that all naval designers are sure that the ship cannot sink. What could possibly go wrong.

Welcome to the lunacy of climate change. It gets worse; far, far worse.

As part of its efforts to get their view of climate history accepted, IPCC scientists waged an extraordinary and ultimately unsuccessful war against substantial evidence that global temperatures were generally higher in the Middle Ages. After much argument the global warming scientists or global warmists as well shall call them, have retreated to the point of agreeing that solar activity accounts for much of observed past changes in climate, albeit while retaining the important caveat that the link breaks down after 1985. A number of global warmers have also recently conceded, from a reading of the current state of oceanic climate cycles, that the world may become cooler over the next few years, before the warming really kicks in. The ocean climate cycles, in turn, are only comparatively recent discoveries and very little is known about them.

There is no reason to think the arguments will end there but despite these shifts, and obvious problems of using projections from assumption-filled computer models piled on top of one another, scientists and economists have piled yet more computer models on top of the temperature projections to prove all sorts of dire results. These models, with their own set of assumptions, have been used to forecast dramatic increases in sea levels, that a million species will become extinct, the world's coral reefs will die out, fish stocks everywhere will collapse, giant storms will destroy beach front housing in rich nations and wash away poor nations altogether, the world's great ice sheets will vanish and agriculture will change out of all recognition.

The assumptions used in these forecasts occasionally conflict. The IPCC projections of CO_2 emissions assume that many poor nations will become richer, per-capita, than America before the end of the century. Well publicised projections for economic damage assume that they will remain poor. But this barely matters, the output of these computer models counts as speculative fiction. Much harsher words could be used.

It gets worse.

Despite doubts about almost all aspects of the science and the output of these endless chain of computer models, activists have been demanding that we spend billions upon billions of dollars remaking whole industrial sectors to reduce emissions. But one of the main solutions proposed, building enormous numbers of wind farms, has never been shown to be of any use. In fact, reports from European countries which have used them in any numbers suggest they are little more than an expensive way not to avoid emissions. Activists have been campaigning for carbon trading schemes, but the only known result of the one, existing scheme worthy of the name, the European Union Emissions Trading Scheme, is to hand billions of dollars to alternative energy projects in the Third World which would probably have gone ahead anyway. The scheme has proved a notable windfall for dam developers in China. No emissions have been saved.

Then there are the strenuous international efforts to get countries to agree to emission targets – efforts which were obviously never going to bear fruit. No one has yet worked out effective ways to stop countries going to war, decide not to pay back international debts, stop drug trafficking, or to prevent them from mass murdering their own citizens, and activists were seriously hoping for a stringent, enforceable agreement on emissions? This was perhaps the deepest level of madness of all, but there is more, much more lunacy. The world's food supply is in peril; we are running out of oil and gas; most of the biosphere is about to collapse; we must reduce the population by more than half; we should start driving electric cars. There is so much that I was simply not able to look at all of it, so I apologise if I have left out the pet hates of some readers.

As part of their efforts to defeat draining criticism, activists and even scientists have compared sceptics of global warming science with those who deny the holocaust. One response is to recall the words of the main instigator of the holocaust, Adolf Hitler. In his book *Mein Kampf* he wrote that the great masses of the people "more readily fall victims to the big lie than the small lie, since they themselves often tell small lies in little matters but would be ashamed to resort to large-scale falsehoods."

In this case we are not talking about a big lie – I'm not suggesting anyone is lying or behaving deceitfully – but if you substitute the word lunacy for lie, you have a good explanation of the fuss over climate change. A lot of people who would not believe little lunacies, because they are used to them, have fallen for a gigantic lunacy. Please read on.

1
FORECASTING FOLLIES

Anyone looking for a cautionary tale about the perils of forecasting, including efforts to make useful statements about climate decades into the future, need look no further than the efforts of solar scientists to forecast the current solar cycle.

The sun has a distinct 11-year cycle marked by spots that astronomers can observe on the face of the sun, which they have been counting and tracking for more than 300 years. At the height of the solar cycle there are lots of these sunspots and the sun is said to be very active, generating plenty of flares and solar storms which affect satellites. At the bottom of the cycle there are few or no spots, and distinctly less solar activity. The sun is quiet. This does not mean the sun becomes noticeable brighter or dimmer, although there are slight changes in its overall energy output. The main change is in the sun's magnetic field and the streams of charged particles, electrons and protons it is always emitting. As magnetic storms and flares can upset satellites, NASA has a particular interest in tracking and forecasting solar activity.

As part of that effort, the space agency declared the last cycle officially over in March 2006, and forecast that the next cycle would be much stronger. Trumpeting a "breakthrough" in solar climate forecasting, using new observations of the solar interior and computer simulations, agency scientists put out a release saying, "Scientists predict the next solar cycle will be up to 30 to 50 per cent stronger than the previous one and up to one year late." (*NASA Aids In Resolving Long Standing Solar Cycle Mystery*, Release 06-087, 6 March

2006.) The document goes into some detail about the model of the internal workings of the sun, including how the twisting of plasma flows creates bursts of magnetic activity.

The sun responded to this interesting explanation of its internal workings by flatly contradicting the forecast. Instead of starting a new, more active solar cycle the sun went completely quiet and stayed that way. Solar cycle number 24, counting from when scientists started tracking the spots, was expected to start in late 2007 or early 2008. As this chapter is being written in early 2010 a few spots from the new cycle have been sighted but unlike normal cycles they have not been building up. One or two at a time appear and move across the sun, but then – nothing.

In May 2009 scientists drew another line in the sand, at least to judge from the graph accompanying the subsequent release (*New solar cycle prediction,* NASA 29 May 2009). The graph suggested that the new cycle will be well underway by the beginning of 2010 and reach a peak sometime in 2013. That release indicates that there were differences of opinion over the 2006 forecasts, which were not reflected in the first release cited, and – with a much needed touch of humility – says that the sun does not pay much attention to human committees. The sun ignored that forecast too.

This is not to drag the name of solar physicists through the mud but to point out how difficult forecasting can be, even with the best computer models and brightest scientific minds at work, and when those bright scientific minds may consider the theory sufficiently well known to attempt forecasting. Now that the sun has flatly contradicted their forecasts, solar scientists will have to go back to the theoretical drawing board. There are indications that the sun may be quiet for a while yet.

After statistically analysing the brightness and magnetic strength of sunspots two researchers at the US National Solar Observatory in Tucson, Arizona, William Livingston and Matthew Penn, believe that there may be deeper processes at work than anything their fellow solar scientists have analysed. Sunspots are highly magnetic regions that are somewhat cooler than the rest of the sun's surface (they appear dark compared to the rest of the sun, but if seen separately would appear very bright). By measuring both the temperature and magnetic strength of the spots over time, the two researchers found that the spots have been warming up and becoming less magnetic. An average of the trend is a descending line which hits the bottom of the graph at 2014. They have concluded that, although sun spots may appear briefly from time to time in the next few years, they will disappear by 2014.

This conclusion is in a paper submitted to the journal *Science* in 2006 but rejected in peer review. That is the journal's editors sent it to other solar scientists who advised the editors that it was not worth the trouble. Sometimes the paper referees may suggest changes, but in this case the paper was simply rejected. With the sun now so quiet the paper has been resurrected from a filing cabinet in the observatory and circulated informally. Livingston told me (by phone from his office in Tucson) that the paper had been rejected on the grounds that it was a purely statistical argument so it would be better to wait and see what happened, and he considered that a fair point. They are now waiting "for the right moment" to resubmit. But what happens after 2014? Livingston says that as they are using a purely statistical argument, without any theory to back it, they do not know.

Aside from being a useful example of forecasting gone wrong, solar activity is thought to play a major part in climate. Another forecast made back in 2000 as part of a theory that solar magnetic activity drives climate also suggests that the sun will remain quiet for some time. We will return to that second forecast in a later chapter. For the moment we should note that one good way to sort out who is right and who is wrong in this debate is to see what the sun does. Readers can look for themselves on NASA's Solar and Heliospheric Observatory or SOHO site. The site always has an up to date image of the sun, with any sunspots labelled, on its home page (http://sohowww.nascom.nasa.gov/home.html). If the sun is still quiet with no sunspots, or just one or two (a last check showed three in a group), when you are reading this then there is a good chance that all the carefully thought out models that formed the solar physics orthodoxy of 2006 missed something big. Just what that big factor may be is a matter for the professionals to sort out. The point is that despite scientists believing that they had the theory right, their forecasts were out by 180 degrees.

An additional, cautionary tale about how forecasts can be very badly wrong can be found in the strident forecasts of American criminologists in the 1980s. As recounted in the book *Freakonomics* (Allen Lane, 2005) by economists Steven D. Levitt and Stephen J. Dubner, when the US crime rate soared in the 1980s various criminologists forecast that crime rates would rise further in the 1990s.

The causes of crime were known in broad terms even 20 years ago. One major factor, for example, is the proportion of the population of young men aged between 16 and 25. If that segment of the population is going to increase with respect to the others then the crime rate should go up. So we would have expected the criminologists to at least be in the right ball park in their forecasts, and it would have been hard to challenge their logic at the time. What other factors could change the result? Instead crime rates fell sharply across the country. Anyone interested in just why these forecasts were also out by 180 degrees should read Levitt and Dubner's entertaining book, bearing in mind that the central thesis of that chapter, that legalising abortion in America cut the number of criminals, is hotly debated.

Nor are forecasting disasters of the magnitude shown in the examples isolated events. They are quite common, even when the theory in a particular area is thought to be settled. They are legion in business and barely worth mentioning in economics, the stock markets – very few foresaw the global financial crisis – social analysis, or in forecasting traffic flows for a new bridge or tunnel. Governments have long given up trying to forecast what markets or industries will do; businesses may gamble on new products or new industries but they have mostly given up efforts to forecast trends. How could anyone have known that water would sell for several dollars a bottle until someone started doing it? IBM in the 1980s would have had to look very hard indeed to foresee the rise and rise of PCs and Microsoft.

Major companies that want to minimise the risk of some unforeseeable trend derailing their business have various options, including constructing a range of scenarios about how their industry will develop. Will mobile phones converge with iPods? Will cameras remain separate devices? – and have a small group watch the market carefully to see which way it develops. Then there are the endless forecasts of matters such as oil prices, the exchange rate and economic conditions in a year's time. Sometimes these are right, sometimes they fail spectacularly. The business sector is well used to the idea of avoiding the risks of forecasting by the use of hedging techniques such as the futures contracts.

Scientists may laugh at this and say that those examples are all to do with business and finance, conveniently forgetting about the solar cycle debacle. Science is different. The planets do not decide that they want their mobile phones to play music or that it is no longer fashionable to orbit the sun; they remain planets obeying long known physical laws. Quite right, but science covers a broad area and natural systems are also not like the solar system. They are usually only partially understood and, like the ocean-atmosphere system, are immensely complicated. Many scientists supporting the climate forecasting orthodoxy seem to be under the impression that climate science is like Quantum Mechanics where the outcomes of experiments have been known to agree with theoretical calculations to several decimal places, or astronomy which can predict the positions of planets thousands of years from now. Nothing could be further from the truth.

A key difference in those examples is the astronomers and physicists have been able to demonstrate that their calculations have some basis in reality. They can point to a track record. Meteorologists can also point to a forecasting track record, particularly as they have a lot of opportunities to make forecasts and then see the next day whether the forecast was right or not. Thanks to satellites and super computers, as well as improved theory, weather bureaus have been making progress.

Whatever the discipline, all the false starts, wrong theories and bad forecasts that led to the physical models used to make those forecasts, have largely been forgotten, except by science historians. One example is the efforts of solar physicists cited above. They have a way to go. Another is the concept called The Ether (or eather) which 19th century physicists swore by, as they thought it was required as a transmission medium for electromagnetic waves among other things. The concept was swept away by relativity. To belabour the point, the only way scientists can tell whether the theory in a particular branch of science is sufficiently settled to make forecasts is when they have used it to construct physical models which have produced useful (scientists say skilful) forecasts.

The Intergovernmental Panel on Climate Change (IPCC) draws on a number of complex computer models known as general circulation models in making its regular reports forecasting the future state of the earth's climate. These reports are gigantic exercises taking several years – the last was in 2007 and the previous one in 2001 – but the GCMs remain at the heart of the exercise. The undoubtedly very advanced and sophisticated GCMs are programmed with an endless number of equations designed to simulate the earth's climate, but they still have to use what scientists know, or think they know, about the interaction of the atmosphere, oceans, ice, winds, clouds, land masses and sun. The problem is that there are several areas of climate, including the behaviour of clouds and the water vapour content of the atmosphere which are crucial but largely unknown. As a result they contain important assumptions, to be discussed later in the book.

Using these GCMs the IPCC reports forecast that earth's climate will become much warmer in coming decades thanks to industrial gases. These forecasts have been taken seriously – so seriously that billions of dollars are being spent on renewable energy projects, carbon trading systems and attempts to reach international agreement on emission limits. This book asks the obvious question. How did we get panicked into spending so much money by the results of computer models that have no forecasting track record and which include a series of assumptions about parts of the system they are trying to simulate?

The vast majority of consumers of climate forecasts will be

unaware that there are any assumptions in these models, including the key assumption concerning humidity. They will not be aware that any sort of forecasting is difficult – let alone forecasting the future state of a natural system. Instead they may be impressed by the fact that the sophisticated models require super computers to run, and that the models use heaps of equations and stuff that they do not understand. Therefore they must be right. In fact, before scientists started using computer models to make climate forecasts such models did not have a high reputation. To put it another way, they had their uses but had to be handled carefully, particularly if the system being investigated was a natural one. Even when they were used in tasks at which such models excelled, such as testing the structure of a bridge where the variables and design components were well know, there could be disasters.

One such disaster was the Millennium Bridge in London which was opened in 2000 to a fanfare of press releases lauding it as "a pure expression of engineering structure". On the opening day thousands of pedestrians steamed across the 320 metre lateral suspension footbridge connecting London's financial district to Bankside, on the other side of the Thames. Very soon the bridge started to sway, then it wobbled alarmingly. It turned out that the bridge designers had tested for every non-human element but had not checked the "natural frequency" of the flexible structure. No one had ever thought to check for such a thing. This natural frequency happened to be close to the frequency of humans walking. A few people walking in step by coincidence in a large crowd started the bridge swaying slightly, then everyone else started walking in step to accommodate the sway. The bridge reopened in 2002 after being fitted with dampers.

A more relevant and much more serious example than that of the bridge, where the only real harm was to the egos of the bridge designers, is the use of models to calculate allowed catch sizes for the Grand Bank fishery off the coast of Newfoundland in Canada. As explained in *Useless Arithmetic: Why Environmental Scientists Can't Predict the Future* (Columbia University Press, 2007) by Orrin H. Pilkey and Linda Pilkey-Jarvis, both geologists specialising in the marine environment, Canadian government scientists constructed elaborate computer models to calculate just how much fish the cod industry could take out of the fishing grounds each year.

The Pilkeys, who are definitely not climate sceptics, say that a computer model to simulate the whole cod fishery is such a complex operation that it is "a virtual impossibility". The complications include: "the interaction of fish with other fish, the roles of predator and prey, the cycle of the food used by larvae and adults, the vagaries of recruitment and mortality rates, the complex food chain, the oceanographic environment in turbulent areas of the ocean where two major ocean currents mix, climatic variations, the habitat loss caused by trawlers and many other such parameters". Little is known about most of those complications. They also say that similar comments can be made about almost any other natural or geological system. They are very complex and you model them at your peril.

Government scientists reduced those complexities by focusing on just one species and put together a computer model. Confident that the fishery was being scientifically investigated and managed, the Canadian government chased away foreign fishing vessels and started to build up the fishing fleets. The politicians expected that the already declining fishing ground would be restored to its former glory. Unfortunately the models were catastrophically wrong and the Canadian politicians did not help by overruling the optimistic catch sizes that were recommended. The fishery closed in 1992 and 40,000 jobs were lost, in the biggest but by no means only such fishing industry disaster. The Pilkeys say that the models were not just wrong in assumptions in one or two factors but wrong in most of them, and the fish stock numbers used in the calculations were completely wrong. The Grand Banks fishery has yet to recover. We will return to the issue of over fishing in the chapter on acid and adaption.

In the book, the Pilkeys are particularly bitter about the use of computer models to forecast what happens when large quantities of sand are dumped on beaches to make them larger and whiter to please the tourists, which happens surprisingly often. They say the models used to forecast the erosion of the added sand are usually wrong, and those using the models suspect that they are inadequate. But the modellers still have to make an estimation to give the finance people a number to plug into the spreadsheets to justify the cost of dumping sand on a beach. The models are not meant to be right, but are there to provide a number.

They briefly mention the climate models favourably, although it is not clear why. (Orin Pilkey has recently been writing articles warning beach front owners of an increase in sea levels of two metres in a century.) But in the book they have harsh words to say about the verification processes used in complex modelling. They say that models, such as simulation of shoreline erosion along a stretch of coast, may be "tweaked" to agree with results for one period but then be found to wrong for another as conditions change. "Perhaps the single most important reason that quantitative predictive mathematical models of natural processes on earth don't work and can't work has to do with ordering complexity. Interactions among the numerous components of a complex system occur in unpredictable and unexpected sequences."

All this means is that despite all that skill, vast scientific effort and expertise that have gone into building climate models, there is no reason to trust them. A creditable forecasting track record would help, but even that is no guarantee. The models may be right perhaps even for the right reasons in a 30 year period, but fail completely in the next 30 years. The computer models that form the basis of climate forecasts have no track record at all. In fact it is worse than that. As we shall also see in later chapters, despite strident claims that the model forecasts are on track or worse than expected, the forecasts that can be checked are either doubtful or wrong.

Global warming scientists defend the models in various ways, and have been defending them constantly since the latest IPCC report released in early 2007, which forecast that global temperatures will increase from somewhere between a mere 0.6 degrees and four degrees over a century. They have also been known to boast that the models are so complex and sophisticated that they must be useful. The problem is that the more complicated they are the more things can go wrong, even if the scientists who built them actually know something about the systems they are trying to simulate. This point is strongly made in one paper by forecasting professionals published in response to the 2007 IPCC report.

The paper 'Global Warming: Forecasts by Scientists versus Scientific Forecasts' by J. Scott Armstrong and Kesten Green is extremely critical of the IPCC report as a piece of forecasting. Armstrong's position as a professor of marketing at the Wharton School, University of Pennsylvania, will not endear himself to the global warming scientists. Green is an economist. Why a professor of marketing? As noted repeatedly, scientists are using the GCMs as forecasting tools and, as shown above, forecasting happens in business all the time. For those interested in forecasting, which is a subject in its own right, Armstrong runs the site www.forecastingprinciples.com.

In the paper posted on the site, Armstrong and Green are particularly hard on chapter eight in the IPCC 2007 report entitled *Climate Models and Their Evaluation* which discusses the use of climate models as forecasting tools. They say the chapter is almost unreadable but the intelligible parts read like a "sales brochure" with variations on the terms "new" and "improved" occurring frequently. After deciphering it, they say it violates the bulk of the known principles of forecasting.

Those principles would take another book, but one is that where there is any uncertainty – and despite all the publicity, the IPCC report is filled with uncertainties – then forecasts should be conservative. That is, the panel should have nominated small increases in global temperatures. The pair are also highly critical of claims that because the models used can simulate temperatures over an historical period, and are very complex, then they must be accurate. Armstrong says that the approach of fitting computer models to past records and then using the models to forecast is known not to work.

After global warming scientists have finished grumbling that Armstrong is a professor of marketing, and that paper has not been peer reviewed, they will object that the 2007 IPCC report is supported by a system of peer review of the results. The comments on peer review show that scientists should not be allowed out without adult supervision. This form of review has merit as a process in scientific journals but is quite useless when applied to forecasts.

As we saw above with the solar forecasting example, peer review is a process undergone by any article submitted to a scientific journal. The editors of the journal send the article off to other scientists in the field, usually two others, for checking. The referees may approve, or suggest changes, or recommend rejection. If they approve the paper is published. This system is the scientific profession's way of reducing the "noise" that can occur when everybody wants to submit papers, particularly as scientific standards can vary greatly. The IPCC also claims to have a system of peer review as draft chapters are shown to scientists who are not the chapter editors for comment, with the implication being that the review process in some way verifies the forecasts. The problem is that even a properly conducted peer review tells us nothing more about the forecast. The review is an assurance that the forecasts are made according to the orthodox theory of the time, whatever that may be, but does not tell us whether the orthodoxy is of any use. There is no substitute for a proper forecasting track record.

In any case, the IPCC's peer review system is not worthy of the name, in part because the comments go to the editors of each of the various chapters that make up the report, and those editors are really in the same position as authors in the ideal peer review system outlined above. Although the editors have various authors writing material for them, they are compiling the chapters and what they say goes.

An analysis of the comments received as part of this review process *Peer review? What peer review?* by climate data analyst and computer scientist John McLean, released in September 2007, shows that chapter editors of the IPCC report faced with unwelcome criticisms from their peers either ignored or rejected them. Available on the Science & Public Policy Institute site (www.scienceandpublicpolicy.org) the paper could only be written after McLean had extracted the comments through the US Freedom of Information legislation. Defenders of the IPCC system insist that the comments have always been public, but this does not seem to be the case. They also attack the SPPI report as not being published in a peer reviewed journal, often throwing in the accusation that the institute must be a front for energy companies (an all-purpose charge).

The IPCC's defenders have not been helped by revelations that at least two pieces of the report, one claiming that 40 per cent of the Amazon rain forest would disappear due to global warming, and another that major glaciers in the Himalayas would melt by 2035, were simply taken from statements by activists. The IPCC has admitted to the glacier claim, but the Amazon claim has since been defended by a prominent scientist. Whatever you make of this, the IPCC's peer review system seems to have major problems. However, the real puzzle is why anyone thought that a peer review process, properly conducted or not, added anything to a set of forecasts in the first place.

As we have seen at its best peer review is one way ensure that published scientific papers add value, so to speak, to a particular scientific disciple. At its worst it is a way to police scientific orthodoxies of the scientific discipline concerned, and by the IPCC as a way of appearing to deal with criticisms without having to change anything. The issue of peer review goes deeper in that pro-warming scientists may occasionally react to pieces of contrary evidence by saying that it has yet to appear in peer reviewed journals. This is a feeble excuse at best. Forecasts are very fragile things in which Murphy's Law (anything that can go wrong…) operates strongly. Scientists serious about making robust forecasts should actively seek out contrary evidence, not use excuses to ignore that evidence. In this and other respects, global warming scientists act as if forecasts of warming are a scientific theory that others have to disprove. They are dealing with forecasts not theories. It's a completely different game.

Not only does peer review tell us nothing whatever about forecasts, it is of questionable value even in the role it is designed to play. As we have seen with the solar cycle example, the process kept out one paper which might well have made a substantial contribution to the debate. There is some evidence that it lets in a lot of papers regarded as breakthrough or highly significant at the time, only later to be proved wrong.

John Ioannidis, an epidemiologist at Ioannina School of Medicine in Greece and adjunct professor at Tufts University School of Medicine in Boston in the US, checked 49 papers in leading scientific journals that had been cited by more than 1,000 other scientists. In other words, they were well regarded papers. But in his subsequent paper in the *Journal of the American Medical Association*, entitled 'Contradicted and initially stronger effects in highly cited clinical research' (J Am Med Assoc, 2005 294: 218-228) he reports that within only a few years, almost a third of the papers had been refuted by other studies.

These flawed papers are not the result of fraud but from more ordinary behaviour such as miscalculation, poor study design or self-serving data analysis. The study concentrated on medical research, which has particular problems in judging the design of studies including the size of the groups being tested, doses of medicine administered and even the analysis of results. All of those factors are subject to judgement of the researchers, and those researchers are under pressure to get published. Although medical research has peculiar difficulties, there is every reason to believe that his work applies to other disciplines.

A subsequent essay in the online journal *Public Library of Science Medicine* (October 2008) co-authored Neil S. Young of the National Institutes of Health in Bethesda, Maryland, Ioannidis and others points out that the top line scientific journals, the ones that all scientists want to get into, are naturally inclined to accept papers with sensational findings. They are also more likely to accept positive results, that is a particular drug has a welcome or unwelcome effect, rather than a study that couldn't find anything although the negative result may be significant.

Ioannidis has since gone on make strong claims about all peer-reviewed papers being wrong, which has been much less well received. But he has also co-authored, the paper 'Persistence of Contradicted Claims in Literature' (*JAMA*, 5 December 2007) which found that "highly cited" observational studies, that is studies that have previously greatly influenced other scientists, persist and continue to be supported in the literature despite strong contradictory evidence.

All this adds up to peer review being a mixed blessing. However, no one has been able to think of a better system and with its help, science tends to go forward rather than back. The problem is that progress can be far from uniform, with different disciplines occasionally marking time for years or decades as scientists defend orthodoxies that may be completely wrong – orthodoxies on which they may have spent a major part of their careers.

One excellent example of this is the saga of the causes of stomach ulcers. In the early 1980s, two Australian scientists Robin Warren and Barry Marshall proposed that a specific bacterium caused stomach ulcers, and not stress and spicy foods as was believed at the time. As is well known they were ridiculed for their pains and their papers were rejected by scientific journals until Marshall experimented on himself by swallowing the bacterium, and developing an ulcer. In 2005 Warren and Marshall were jointly awarded the Nobel prize for medicine. So the orthodoxy was overcome and science moved forward, but it took time.[1]

Another, much larger example of how highly trained scientists can cling to an orthodoxy for decades, despite considerable contrary evidence, is that of Freudian psychiatry. For a time at the beginning of the 20th century Freudian psychoanalysis was wildly popular, even

---

1 A friend of mine who is PhDing in astronomy at Monash University in Melbourne often tells me about the shortcomings of the disk model of the formation of the solar system, and he may well have a point. When the Voyager satellites sent back data on Saturn and Uranus in the 1980s a good part of what they observed surprised scientists. Occasionally one reads about efforts by scientists to simulate the formation of the solar system and, after making extreme assumptions, manage to get something that looks like a solar system, give or take a planet. This set of brave computer modellers then have to make reality fit their computer model. So why is, say, Mercury missing? We are then solemnly informed that the planet much have come from somewhere else – that is, it wandered in to be captured by the sun after the solar system proper was formed. The fact that the model of the formation of the solar system might be completely wrong does not seem to be considered. Rock samples from the planets might confirm, or refute, this egregious use of computer modelling but those rocks are not going to be collected any time soon. In the meantime the orthodoxy stands.

spawning the sub-discipline of psychohistory with its own learned journals, and influencing other disciplines such as anthropology. Despite all this strenuous academic activity, it took psychiatrists several decades to realise that what was new in Freud's theories was of no use in treating patients, and what was useful was not new to Freud. He did not invent the concept of the subconscious or the approach of talking to patients, but popularised the notions.

The real surprise is not that these theories have been abandoned but that they lasted so long. As various writers have pointed out, Freud essentially made up his theories about infant sexuality and hidden memories from a few observations and actively resisted suggestions that they be subjected to clinical trials. (See *The Decline and Fall of the Freudian Empire*, Hans Eysenck, Viking, 1985. A more general overview of the now passé sport of Freud bashing can be found in *Great Feuds in Medicine*, Hal Hellman, John Wiley, 2001.) Other writers have tried to salvage Freud's reputation by claiming, among other things, that his work marked a major change in the way the mentally ill were treated, which is at least arguable. There are also still analysts, particularly in America where being analysed is something of a cult, and Freudian notions are often used in films and books, but his ideas have been tossed aside by scientists.

So much for scientists as paragons of wisdom who cannot be collectively wrong. They are capable of shocking misjudgements, both individually and collectively, and of defending those misjudgements for decades in the teeth of considerable evidence to the contrary. In other words, they can be just like the rest of us, although science does eventually go forward rather than back.

Defenders of the IPCC forecasts will use the analogy of consulting with a heart specialist. If such a specialist tells you there is trouble then you should listen; so if climate specialists tell you something is wrong then you should take his or her word for it. The analogy is wrong. You should listen to heart specialists because they can point to a track record of medical interventions that improved patient health, rather than made it worse. Climate specialists on the other hand have no such track record, and are essentially making pronouncements about a system of which they have very little knowledge, as we shall see.

Obviously a lot of scientists disagree with that assessment and some of them are very distinguished, so surely that should count for something? Climate activists frequently pump up this line of argument into claims that almost all scientists approve of the IPCC forecasts, and that the sceptics are a tiny, cranky minority with no credibility. This is a wild overstatement of the case. The main claim that increasing amounts of carbon dioxide due to industrial activity are warming the earth amounts to a scientific orthodoxy, perhaps a strongly held scientific orthodoxy, but no more than that. Orthodoxies don't count for very much unless, to labour the point, they are shown to be useful in some way.

The more prominent critics of the IPCC line include Richard Lindzen, a professor of Meteorology at Massachusetts in the US; William Gray, Emeritus Professor of Atmospheric Physics at Colorado State University; and Roger Pielke Jr, a professor in the environmental studies program at the University of Colorado. Roy W. Spencer, principal research scientist at the University of Alabama in Huntsville and a formerly a senior scientist for climate studies at NASA, recently wrote a book *Climate Confusion* (Encounter Books, 2008; available on Amazon.)

In Australia prominent critics include Bob Carter, an adjunct professor at the Marine Geophysical Laboratory at James Cook University in Queensland; Stewart Franks, an associate professor at the University of Newcastle school of engineering specialising in the environment; David Evans who helped build the carbon accounting models for the Australian government to track carbon in plants, debris, solids and agricultural products; and William Kininmonth, former head of the National Climate Centre at the Bureau of Meteorology, and a former Australian delegate to the World Meteorological Organisation's Commission for Climatology.

Garth W. Paltridge, an atmospheric physicist and former director of the Institute of Antarctic and Southern Ocean Studies, among other posts, has also written a book *The Climate Caper* (Connor Court, 2009). On the policy side, and cataloguing some bizarre behaviour on the part of environmental scientists, a professor and head of the school of government at the University of Tasmania, Aynsley Kellow, has written *Science and Public Policy: The Virtuous Corruption of Virtual Environmental Science* (Edward Elgar, 2007). Perhaps the most controversial is Ian Plimer, a professor of mining geology at Adelaide University, who recently strongly attacked global warming science in his *Heaven + Earth* (Connor Court, 2009).

For those still not convinced by any of this there is the Manhattan Declaration, the list for which can be found on the site www.climatescienceinternational.org. This is list of more than 100 scientists who managed to get to Times Square in New York on a particular day in March 2008. There are several such lists, including another to be found on the US Senate Committee for the environment and public works site headed, 'Over 400 Prominent Scientists Disputed Man-Made Global Warming Claims in 2007'. *Scientists Debunk "Consensus"*. The report names eminently qualified scientists including noble prize winners, complete with a quote from each scientist on some aspect of their opposition to global warming science.

The usual reaction of activists faced with irrefutable proof that scientists do not overwhelmingly agree with their position on carbon dioxide is to search the list for a scientist who may have, say, once worked for a oil company, then triumphantly announce the result. There you are, that scientist has worked for an oil company, and this other scientist said something doubtful about acid rain 20 years ago, or whatever, therefore the whole list is a front for the oil and coal companies. Whether the mainstream media believes those excuses or not, the existence of such lists is rarely reported.

As for the claims that 2,500, sometimes 1,500, of the world's top climate scientists, support the IPCC reports, those figures have no basis beyond the panel's press releases. The analysis by John McLean cited above, points out that the critical chapter in the 2007 report is chapter nine, which attributes global warming to human activity. Everything else depends on those forecasts. That chapter attracted comments from just 55 individuals in the peer review process, plus another eight comments from the government representatives. Of those making comments, 31 could be said to be involved in the report already in some way as editors or as having papers cited. There are many other scientists involved in the report but they are concerned with areas such as mitigating or adapting to climate change, and simply use the projections produced in the climate chapter report.

Climate activists will, of course, point to their own lists of eminent scientists who say that human activities are warming the earth, and those lists certainly include many prominent individuals. They can also point to long lists of science academies around the world, such as the US Academy of Sciences and the Australian Academy of Sciences which have endorsed induced global warming. A response to this is that the academies never polled their members before expressing their support and, as we shall see in the chapter on acid oceans, these academies will endorse the most extraordinary statements about almost anything provided it means that their members will get more research funds. With honourable exceptions their actions are more like those of trade unions than independent advisory bodies.

In any case, before we get further into counting up scientists for and against human-induced global warming, the whole issue is essentially irrelevant. The real question has never been whether scientists agree to a particular orthodoxy, but whether the orthodoxy is right. As we have seen, and shall see again, one has little bearing on the other. Instead, as used to be the case, forecasts should be a test of the orthodoxy. Scientists should make forecasts and, if they get something right, congratulate themselves. If the forecast proves wrong they go back to theoretical drawing board.

Surely scientists or experts in a particular field armed with complex computer models are more likely to be right than, say, an interested amateur, or a monkey throwing darts at a list of possibilities? Actually no. Forecasting gurus Armstrong and Green in the paper cited above, point to research in politics, economics and areas such as long term electricity consumption which indicates trained professionals do little better than ordinary citizens in forecasting. They can also point to long lists of pronouncement by scientists which have been proved completely wrong by events. In 1932 Albert Einstein famously declared that nuclear energy would never be obtainable, and in 1956 the famous physicist John von Neumann declared that in a few decades energy would be free.

A more relevant and better known example is that of scientists in the 1970s repeatedly and stridently warning all concerned that the earth was about to plunge into another ice age. As noted in the book *The Real Global Warming Disaster* by UK journalist Christopher Booker (Continuum, 2009), George Kukla, a Columbia University astrophysicist, and Robert Matthews, head of the Geological sciences department at Brown University, Rhode Island, and others reacted to a decline in global temperatures of the time to convene a conference of scientists from America and Europe. They were so alarmed by the conference results that in December 1972 they wrote to President Nixon to warn him of the looming ice age. The letter had barely reached the president's desk before temperatures started increasing.

The global cooling scare never had the same appeal as global warming as it could not be blamed directly on industrial emissions, and so did not jibe with the innate beliefs of activists who are convinced that humans are guilty of something. But it does show that when scientists get together that does not mean their forecasts are more likely to be right. All it means is that they can be wrong or right collectively.

Armstrong and Green, in the paper cited above, also point out that when it comes to forecasting the use of computer models makes no real difference to this bleak picture. The two men say that the only way known to make forecasts more accurate is to give the forecasts feedback. "When forecasters get substantial amounts of well-summarised feedback about the accuracy of their forecasts and about the reasons why the forecasts were or were not accurate, they can improve their forecasts."

As we have seen, there is nothing in the history of forecasting or the use of computer models to suggest that we should pay attention to them, simply because a number of scientists say that they are right. Instead, there are lots of reasons for thinking that climate forecasters have fallen into the classic trap of failing to looking beyond the concerns of the moment, which in this case are industrial emissions.

One classic text in forecasting is that of Steven P. Schnaars, *Megamistakes Forecasting and the Myth of Rapid Technological Change* (The Free Press, New York, 1989) which examines the period of the 1960s through to the 1980s when scientists and experts of all sorts freely made predictions. Many of these forecasts were about how quickly new technologies would be adopted, with a major source of error being the assumption that consumers would rush out and buy new technology just because it is new. In fact, technologies such as the fax machine and the early VCRs took time, and had to be improved considerably in performance and price, before being accepted by consumers. Other forecasts, such as the future use of gas and nuclear energy simply reflected industry conditions at the time the forecasts were made. Schnaars, another professor of marketing, is dismissive of the far more primitive computer modelling of the time, saying that common sense plus a calculator often yielded far better results.

He emerged with one conclusion, that forecasts often simply reflected whatever the forecasters happened to be concerned about at the time. As a result they often tell historians far more about the concerns of the people who made the forecasts than say anything useful about the future. The laboriously constructed IPCC forecasts based on far from complete science (as we shall see in the next chapter) and without any forecasting track record to speak of, are unlikely to break Schnaars' rule.

2
ICE AGE ESCAPADES

One frequently repeated statement in the ongoing debate about industrial emissions being responsible for part of recent change in climate is that the science is settled. But even the briefest of glances at this area by any lay person able to read or look at graphs shows that this claim is simply wrong. The earth's climate has been known to vary considerably in the past, both in to and out of ice ages and with much smaller variations in-between. Some of those variations are known to have involved temperatures higher than modern times. Scientists can explain bits and pieces of these variations, and parts of that theory involves carbon dioxide, but there are problems – lots of them. In addition, as we shall see in another chapter, attempts to explain away the earth's stubborn behaviour not to heat up in the last decade or so by different global warming groups indicates that they cannot agree among themselves on even seemingly basic points about the science.

When confronted with these problems scientists wave them away saying that the simple laypeople/journalists asking those questions have no idea of the science, but they (the scientists) do and they know that carbon dioxide causes climate change. The models prove it, so take their word for it. Well we won't.

Let's start with some basics. Figure one shows temperature changes over the past 420,000 years.

Ice Ages and Intergalacials

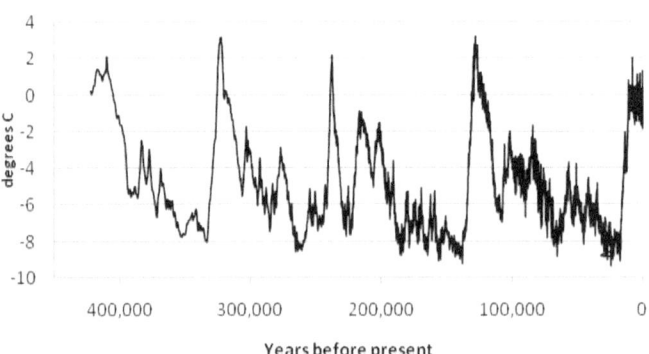

Figure 1: Temperatures over the most recent 420,000 years of earth's history. Note that the graph time line is from left to right. All of recorded human history is in a couple of squiggles on the far right. Source: Vostok cores and Petit et al, 1999.

As you can see the temperature changes measured from ice in ice cores at a place called Vostok in Antarctica, where the ice has built undisturbed for all that time, show a distinct pattern of long ice age and brief, warm interglacial (the period between two ice ages) such as the one we are in now. The data is publicly available if you search from the US-based Carbon Dioxide Information Analysis Centre (http://cdiac.ornl.gov). To get this data scientists drill into the ice to take very long columns of ice which they cart back to laboratories and analyse slice by slice. The further down the ice core they go, which may be kilometres long, the further back in time the ice was laid down. They also analyse the air trapped in bubbles in the ice. By measuring very fine details in the ice in this long-trapped air such as the ratios between different isotopes of oxygen and hydrogen (see box at end), they can estimate temperature at the time the ice was laid down as a season of snow. They also analyse carbon dioxide levels in bubbles in the ice. There is a great deal more to it, enough for another whole book on discussing the procedures and on calibrating the results so that they can be placed accurately on a timeline. The above graph, incidentally, stops about 5,000 years ago.

There are other ice cores and other ways of measuring past climates such as taking similar columns of sediments laid down by the sea or lakes and looking for changes in the composition. There is a whole science of analysing the ancient pollen trapped in sediment. Limestone formations in caves can be used to track climate and for the last 10,000 years or so and scientists can track tree rings, which is another complex branch of science all to itself. Rings in living trees may go back a few centuries, but fossilised trees and bits of wood have been found and analysed to go much further back. Bore holes can also be used to reconstruct temperatures over centuries. All this painstakingly collected data are called temperature proxies; that is they stand in for instrument readings from ancient times. However, the data from the Vostok ice cores is the most widely used and analysed for it shows a connected account of temperatures for the broad sweep of the Earth's last four ice ages. Further it can be connected to a graph of CO_2 concentrations in the air, which is not shown.

The actual temperature figures from the graph do not mean anything. The zero degree line has nothing to do with the freezing point of water, but is a handy reference point. All temperatures are expressed as variations from that line, and we need only concern ourselves with the changes. Waving away those irritating technicalities for the moment, we are big picture people after all, you can see from the graph the previously mentioned cycle of ice ages broken by brief interglacial periods, including our own interglacial on the right. The whole glacial-interglacial cycle takes about 100,000 years to complete each time. Humans or Homo Sapiens evolved about 150,000 years ago before the previous interglacial, reaching Australia 40,000 years ago (occasionally the figure is given as 60,000 years).

With so much water locked up in enormous ice sheets, sea levels were up to 40 metres lower than modern times at the depth of the last ice age, so migrating to Australia via Asia was a matter of walking combined with short boat trips. For unknown reasons it is generally thought that humans took much longer to reach the Americas over the dry land that is now the Bering Strait. Once at the straits, the first Americans may have found the way blocked by two enormous ice sheets, the Laurentide which stretched across Canada down to well past New York, and the Cordilleran in the Alaskan sea. Around 13,000 years ago, as temperatures were increasing, the two sheets melted sufficiently for small bands of hunter-gatherers to emigrate through the gap over many generations. That is one theory but there is plenty of controversy over this with others putting American settlement at up to 40,000 years ago.

The early humans adapted to the much colder temperatures in many ways. About 12,000 years ago early Europeans lived for months deep in caves, emerging to hunt in the brief summers. When temperatures increased they abandoned the caves, then they invented agriculture and, a little later, mobile phones. All of recorded human history occupies a couple of squiggles at the very end of the line, or would have if I had added other work by paleoclimatologists (the study of past climates by examining the proxy records mentioned) on more recent temperatures. The Vostok ice core records end 5,000 years ago.

This distinct 100,000 cycle is known to have gone on for about a million years. For the two million years before that the earth went through a similar repeating ice age-interglacial cycle, but with a period of 40,000 years and in generally warmer conditions The ice ages were not as cold and the interglacials warmer. The whole three million year period is, in turn, part of an ice-age period for earth in which temperatures have cooled. There have also been hot house periods as part of a host of major variations in earth's geological history, but the fine detail of all those variations is missing. Scientists know in general they occurred but have to infer changes from data points that may be a million years and more apart. At least with the Vostok cores scientists have been able to construct a connected account of the both temperatures and CO_2 concentrations for the past 420,000 years.

As noted scientists can construct a graph of CO_2 concentrations, expressed as parts per million (10,000ppm is 1 per cent so 330ppm is 0.033 per cent of the atmosphere), which is not shown here but is very similar in that it also shows a distinct 100,000 year cycle. So isn't that proof that CO_2 affects temperature, even in the very low concentrations measured? Ice cores taken from Vostok a few years before the one graphed above showed that temperatures and CO_2 moved in lock step, so it was assumed that CO_2 increases were driving those of temperatures. The results from those earlier cores were highlighted in the 2006 documentary *An Inconvenient Truth*, presented by former US vice president Al Gore, which has arguably helped create much of the present obsession with carbon.

The problem is that later cores and better analysis of the ice showed the sequence in more detail and, as is widely acknowledged, the CO_2 changes lag those of temperature. This point has been explored through a series of scientific papers with the latest published in the journal *Science* last year by a team led by Lowell Scott of the Department of Earth Sciences, University of Southern California. Entitled 'Southern Hemisphere and Deep-Sea Warming Led Deglacial Atmospheric CO_2 Rise and Tropic Warming', the paper states that between 19,000 and 17,000 years before the present – that is, at the beginning of the end of the last ice age – deep sea temperatures warmed by about two degrees C about 1,000 years before carbon dioxide concentrations starting increasing. Other papers give different time lags, but nothing less than centuries.

The Al Gore documentary was made after the results of the later Vostok cores became known, but the producers and writers did not bother to mention this complication – a point made many times now but always brushed aside or ignored by warmers. In any event global warming scientists have been able to come up with an explanation for the temperature changes involving CO_2 that is at least plausible.

So why does the earth's climate change, and why does it change on such a regular schedule? The current orthodoxy is that the trigger for the end of the ice ages are periodic, slight changes in the earth's orbit and the angle of the earth's pole. The theory was first proposed by James Croll, a self-taught 19th century physicist and greatly extended by a Serbian civil engineer and geophysicist Milutin Milankovitch who died in 1958. The theory, which has fallen in and out of favour a couple of times since it was first proposed, involves three sets of variations. First the earth's orbit varies slightly in a 100,000 year cycle. Then the angle of the earth's North South pole varies. The earth does not sit straight up, parallel with the sun, as it orbits but is on a slight angle, with that angle resulting in our seasons. When the Northern Hemisphere is closer to the sun thanks to the angle, then it is summer in the North. This angle becomes smaller then larger again, on a 41,000 year cycle. In addition, the North South pole will move in a circle on a 25,800 year cycle. Just think of a top wobbling (the technical term is precessing). Those three cycles do not change the amount of overall energy the earth gets from the sun but will change the strength of the seasons. Summers may become hotter or colder, winters may become colder or milder.

The changes in solar heating in different seasons are not enough to end ice ages by themselves, as the effects are generally weak. But, scientists suppose, hotter summers in the Northern hemisphere which contains most of the land will melt enough ice to expose bare rock. Ice reflects sunlight but rock absorbs it to become warmer, and that extra warmth may melt more ice. There is a positive feedback. Whatever the mechanism, the additional warming is supposed to be crucial. As is generally accepted once the warming is underway it warms the oceans, and warmer oceans hold less carbon dioxide so there is more in the atmosphere. As we have seen the warming goes on for a long time before CO_2 concentrations shift appreciably, but global warmers say that when the extra CO_2 does arrive it takes over to become the main driving force. The extra CO_2 warms the earth more, and that extra warmth drives more CO_2 out of the ocean and so on. Although a 1,000 years and two degrees seems like a lot of time and warmth to explain away this story is at least plausible, particularly as shifts in comparatively low concentrations of CO_2 do have a warming effect as we shall see in a another chapter. So let us move on to further complications.

Another glance at the graph indicates that temperatures in the previous two interglacials seem to spike to well above modern temperatures, but the ice core readings for the time show that CO_2 levels were at about 300 parts per million or far less than present levels. (There is some argument about the ice core results for CO_2 which we shall discuss in another chapter.) *The Arctic Climate Impact Assessment Report 2005*, compiled by a group called the International Arctic Science Committee and used as a source for the 2007 IPCC report says that during the inter-glacial preceding this one, known as the Eemian, "the winter sea ice in Bering strait was at least 800 kilometres further North". In addition, "lake pollen records show deciduous forests (characteristic of warm conditions) across much of Western Europe". The 2007 IPCC report says that polar temperatures at the time were 3 to 5 degrees higher than now, but attributes the difference to Milankovitch cycles.

While we are on the subject another problem is that if initial warming is supposed to push more carbon dioxide into the atmosphere which then creates more warming and so on, and CO_2 is supposed to be such an effective greenhouse gas, why didn't temperatures and CO_2 levels just keep on going? Why did they stop climbing and reverse, particularly given that the astronomical cycles just accentuate the seasons rather than add to warming overall? One vague explanation is that perhaps the extra warming melts ice and the additional cold water causes cooling. Perhaps there is a tipping point? These speculations seem to indicate that we would need to worry more about the earth being tipped into an ice age than warmer temperatures, but let us move on. There are also instances where CO_2 levels don't seem to be connected to temperatures at all. About 130,000 years ago, CO_2 levels soldier on for many thousands of years despite the earth falling into an ice age.

Milankovitch cycles have also run into distinct timing problems. On the surface it looks good. One cycle is 100,000 years long which would seem to fit in nicely with the last million year's worth of ice age-interglacial cycles noted above. One peak in summer warming in the Northern hemisphere occurred about the time of the Eemian and another occurred just a few thousand years ago, or in the lead up to the time when temperatures were known to be much warmer in the middle of our own interglacial. Our interglacial is called the Holocene, so the period is known as the mid-Holocene maximum.

One problem is that the 100,000 cycle is supposed to be weakest, so why it so important compared to the other two and why the switch from a 40,000 year cycle? A 400,000 year cycle – a combination of the others – is supposed to be the strongest of them all, but that does not seem to show up at all in the geological cycle. For that matter why is the earth in an ice age phase with generally cooling temperatures? More research has resulted in more questions.

In 1992 measurements of climate with precise dates became available from a rock formation in a water filled cave in the US state of Nevada called Devils Hole. From measurements of calcite deposits in the walls of the cave, scientists were able to track the variations in the isotope of Oxygen with an atomic weight of 18 (see box at end) over hundreds of thousands of years. Changes in that isotope are linked to temperature. At the same time they were able to match those results with dates calculated by measuring levels of radioactive isotopes of Uranium and Thorium also present in the calcite. In other words they were able to derive an independent record of both dates and temperatures from the same sediments. Those sequences did not fit the Milankovitch cycles. The comparison shows that the rise in temperatures that led to the Eemian happened thousands of years before any warming from the cycles could have taken effect. In fact, global temperatures started to rise when the solar warming effect from these cycles was at its lowest rather than its highest. As scientists have found the Milankovitch cycles quite useful they are not going to abandon the theory over results from a single, water-filled hole. But the problem has since been confirmed by analysis of sediments in the Bahamas and Corals in Papua New Guinea, showing that warming was well under way before the cycles could have taken effect.

These arguments are set out in the book *Ice Ages and Astronomical Causes* (Praxis, 2000) by Richard A. Muller and Gordon J. MacDonald. Muller is a professor of physics at the University of California and the late MacDonald was the director of the International Institute for Applied Systems Analysis in Austria. The book points to various major problems with Milankovitch cycles, and says that part of the reason they have become so widely accepted is the practice of "tuning". Researchers faced with long sequences of climate records from sediments will match the shift out of ice ages shown in their results to the cycles, which can be precisely dated. In other words, researchers have been assuming the Milankovitch cycle theory is correct when working out dates for their own data, then using the results to "prove" the cycle theory. This tuning problem has been known to happen in other disciplines. A weird part about all this, as the scientists admit, is that the Milankovitch cycle patterns are an excellent match for the change in and out of ice ages, but just a few thousand years too early.

Muller and MacDonald put forward their own solution to the 100,000 cycle problem, and why the earth switched from a 40,000 year to a 100,000 year cycle. This involves another cycle where the earth moves through the orbital plane every 100,000 years or so and collects interplanetary dust resulting in cloud formation. The theory is interesting and perhaps not as far-fetched as it sounds on the surface, but both it and the Milankovitch cycle theory also have to hold off the up and coming, and more comprehensive, solar magnetic theory to be discussed in a later chapter.

For now we should note that scientific grumbling over the cycle theory is becoming louder and that Muller and MacDonald are by no means alone, at least in pointing out that the existing theory has major problems. To date scientists have concentrated on renovating the cycle theory so that it fits the data, as well as explain why the ice age – interglacial cycle suddenly switched from 41,000 years to 100,000 years, which is proving to be a major problem. Perhaps the key is solar warming in the southern hemisphere? Perhaps different part of the cycles are important and summer temperatures are not the key? There is a lot of debate about this and the Devils Hole results in the scientific journals. Scientists will get back to us.

As you can see from the graph the Holocene is already longer than the preceding three interglacial periods at least. Estimating from graphs that are better than mine, the Eemian lasted less than 10,000 years and the previous two interglacials lasted perhaps just 5,000 years. The Holocene, in contrast, has clocked up 10,000 years plus. Further, those 10,000 years have remained comparatively warm, in contrast to the previous interglacials in which temperatures spiked then fell away again very quickly. Temperatures have spiked for this interglacial, in the already mentioned mid-Holocene maximum, but the subsequent decline in temperatures has been much more gradual. No accepted theory accounts for any of this.

In all cases the final collapse into an ice age was abrupt to say the least and, again, there is no real theory to explain the collapse, apart from vague references to the cycles. The atmosphere-oceanic system just seems to trip over itself to fall 10 degrees and more. Also, as you can see from the graph, once the system falls into an ice age it tries to lift itself up several times in periods called interstadials (which don't fit into the orbital theory), but otherwise becomes colder and colder, until finally it kick starts properly again. As modern global temperatures are still comparatively high such a collapse is not likely to happen any time soon, but with climate theory in its present state there is no way to predict when it will occur. Orbital cycles theory suggests another 14,000 or so, although it doesn't have much to say about the other interglacials ending. The solar magnetic theory, which is a better fit but still counts as just a theory, says another 10,000 years. When the collapse does occur, it may take less than a couple of centuries.

When the earth is warming up out of an ice age, it seems that it can take major shocks and still keep warming. As is hopefully just visible on the graph there is a major dip and bounce back in temperatures about 8,000 years ago called the Younger Dryas event. This is attributed to a sudden out-flow of water from the melting Laurentide ice sheet which still covered most of Canada at the time. That influx of cold water shut down the gulf stream, which is a very important part of the earth's ocean circulation system. A scenario like that is proposed as the basis for the 2004 Hollywood blockbuster *The Day After Tomorrow* which depicted global freezing as a result of global warming. The Younger Dryas event took generations and not days as in the film, and the results were not anything like as bad, particularly as the film combined freezing conditions with much higher sea levels. As already noted, sea levels fall during ice ages. However, the film was entertaining.

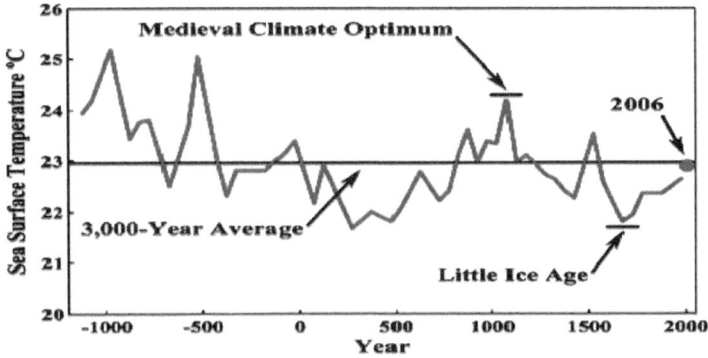

Figure 2: temperature cycles in the past three thousand years using sea surface temperatures as a guide. Courtesy of the NIPCC (Non-government IPCC) 2009 report.

Once the earth climbed out of the ice age and melting ice sheets no longer upset climate there were major variations in temperatures over the Holocene which show a distinct cycle, known as the Dansgaard-Oeschger cycles after the scientists who discovered it. These cycles, some of which are shown in figure two, have been confirmed any number of ways and have been discussed at length in a book by S. Fred Singer, a climate physicist and research professor at George Mason University, who is also a leading sceptic (he founded the NIPCC mentioned in the caption) and Dennis T. Avery, a senior fellow at the Hudson Institute in the US. In the book entitled *Unstoppable Global Warming* (Rowman & Littlefield, 2007), they put its length at 1,500 years, give or take 500 years, and say that it can be traced back through one million years – a total of 600 cycles, which is an average closer to 1,700 years. They also discuss a mechanism for this cycle which we will examine in its place. For the moment all we really need to note is that there is no real dispute that a cycle of some sort exists.

These cycles have left their mark on history. There is some historical evidence that temperatures were generally warmer in Roman times and colder during the dark ages, then higher in Medieval times and colder in the last few centuries. The two most recent climate episodes are known respectively as the Medieval Warm Period (MWP), or sometimes the Medieval Climate Optimum (or anomaly), and the Little Ice Age (LIA). The MWP is thought to have lasted from about 800 CE to perhaps 1300, albeit with different start and end dates in different continents. America was badly affected by droughts, and there were frequent climate shifts. The Little Ice Age, which succeeded the MWP everywhere and included major variations in temperatures, lasted from about 1300 to perhaps 1850 which is about when the world seriously started to industrialise. Modern weather records start in the 18th century with the invention of the thermometer and barometer, but those were only used in certain areas. Anyone suddenly transported back to the Little Ice Age would not notice much difference day to day, incidentally. They would simply think that the summers were generally cold and the winters freezing, but there would still be plenty of variations in the seasons. Figure four is a simplified version of the MWP-LIA sequence.

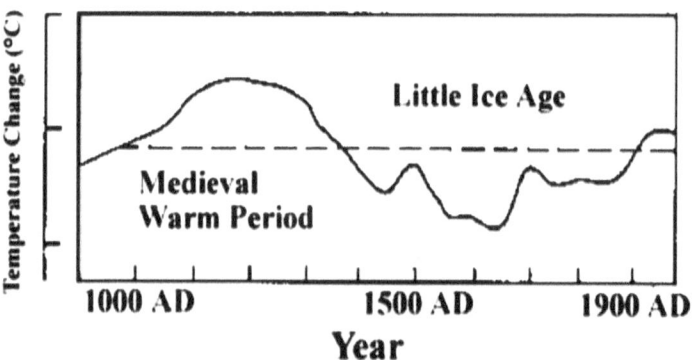

Figure three: the Medieval Warming Period and the Little Ice Age according to the IPCC 1990 report. This was before the panel became both famous and convinced of its mission. The graph is from earlier climate work but should give you the general idea.

So here we have the problem. We know temperatures started to drift up before industrial emissions could have been said to be a problem and that we are very likely in the high part of one of these cycles and that is about it, so far. Modern warming will be discussed further, but no one now is really arguing that at least part of the increases in global temperature since the Little Ice Age is natural. The argument over climate change is about how much human activity has added to the present warm climate, if anything at all. You may have seen newspaper reports stating that, say, seven of the last 10 years are the warmest on record. This means the warmest on the instrument record, but as proper record keeping using instruments did not start until the 18th century, near the end of Little Ice Age, this statement tells us nothing.

Not shown in the graphs above is a distinctly warmer period in the middle of the Holocene. The *Arctic Climate Impact Assessment Report* cited above notes that "over most of Russia, forests advanced to or near the current Arctic coastline" between nine and seven thousand years before the present, and retreated to their present position by between three and four thousand years before present. "Evidence for a mid-Holocene thermal maximum in Scandinavia is considerable", with tree lines on mountains up to 300 metres higher than present. The present orthodoxy is that it is due to a Milankovitch cycle boosting the ocean-atmosphere system with higher summer temperatures in the Northern Hemisphere (presumably overcoming lower temperatures in the Southern Hemisphere). The cycle is supposed to have peaked a couple of thousand years or so before the maximum, but the difference is put down to the extra heat taking time to work its way through the system. However, as noted above, the cycle theory has run into distinct timing issues for other interglacials which indicate that the cycles may have much less effect than previously thought.

Quibbles over the cycle theory aside, global warming scientists could blame part of the temperature variations and the mid-Holocene maximum on orbital cycles, but it was not as easy to get rid of the Medieval Warming Period or Little Ice Age. There is plenty of historical evidence that that the MWP was warmer overall than present times in certain areas at least. One rough measure of temperature is known to be the increase and decline of glaciers. As we are told endlessly glaciers have been shrinking. That's right, overall glaciers have been shrinking because climate has been getting warmer over the past century, although there is considerable variation due to regional conditions. By carefully examining the area around glaciers scientists can trace their rise and decline as a proxy for the cycles mentioned above. A wealth of data on the rise and decline of glaciers is contained in papers in scientific journals such as *The Holocene* and *Climate of the Past*.

We will discuss ice a little more in another chapter, but scare stories about glaciers vanishing virtually overnight, in geological terms, have been found to be just scare stories. Glaciers overall are receding, because temperatures are generally higher than they were in the Little Ice Age. The key questions are whether any of the temperature change is the result of human activities, and whether glaciers have, on average, melted past the low point of the MWP.

There is other evidence that temperatures were higher in the MWP. It is known from the Doomsday Book, a sort of medieval tax database, that there were vineyards in England in medieval times, which is hardly possible at modern temperatures. In his classic work on climate, *Climate History and the Modern World* (Routledge, second edition 1995), the late H. H. Lamb, formerly emeritus professor in the School of Environmental Sciences at the University of East Anglia, estimates from evidence of the agricultural practices of the time, that average summer temperatures in England in medieval times were probably between 0.7 and 1.0 degrees warmer than the present day. In central Europe they were between 1.0 and 1.4 degrees higher. There is also evidence that frosts were not as severe.

In 2001 three scientists, A. Hiller and T. Boettger of the UFZ Centre for Environmental Research in Halle, Germany, and C. Kremenetski of the Institute of Geography, Russian Academy of Sciences, took samples of fossilised wood from mountains on the Kola Peninsula in north-western Russia. They concluded that at around 1000 to 1300 AD, that is around the "Medieval climatic optimum" pine forests grew at least 100-140m above the modern pine tree-line ('climatic warming recorded by radiocarbon dated alpine tree-line shift on the Kola Peninsula, Russia', *The Holocene*, 2001). This implies that temperatures in the region were then much higher than present.

Besides all those dry results from analysing fossilised trees, looking for ancient tree lines and working out just how high up in the hills farmers planted crops 1,000 years ago, various commentators have also noted that the many of Europe's major cathedrals were built in the medieval period, and Viking settlements in Greenland thrived. It was a time of optimism and increasing wealth. As we shall see in a later chapter, the peasants still led very hard lives but at least harvests were mostly regular and predictable.

All of this evidence caused some angst in the global warming camp. For a time they managed to get rid of it in a strange episode known as the hockey stick graph. In 1998 Michael E. Mann, a climatologist and associate professor at the Pennsylvania State University, along with Raymond S. Bradley and Malcolm S. Hughes, published a statistical analysis – a compilation - of a range of the temperature proxies such as tree rings, spliced with modern temperature records.

Such a compilation is not straightforward. Although scientists have become very good at analysing ice or reading the fine detail of sediment laid down on ancient shore lines or tree rings as temperature proxies, the results are still far less accurate than modern instruments and there may be just a few data points as opposed many readings from modern instrument networks. Scientists say that the data has low resolution. Another problem is that different ways of measuring temperatures – such as tree rings and analysing ocean sediment – give different absolute temperatures (they may show similar trends but be in different parts of the same temperature graph). In other words, mixing temperature proxies into a coherent temperature record and combining it with high resolution modern data, is an art not a science. This lack of precision, incidentally, also means that it is hard to point to temperature changes in recent times and say that the change is faster, or slower, than anything that has occurred previously, particularly given the slight change involved.

Mann and his colleagues dived into the pool of proxy temperature statistics to emerge with an analysis that does not show any MWP-Little Ice Age pattern but does show a major, sudden leap in temperatures in modern times to well above that of any climate in the past 1,000 years ('Global-Scale Temperature Patterns and Climate Forcing Over the Past Six Centuries', *Nature*, 1998.) The shape of the graph was similar to that of a hockey stick.

The MWP-LIA pattern was quite well known at the time, and had even been confirmed only shortly before Mann's paper by an analysis of 6,000 bore hole results in the database of the International Heat Flow Commission. For its own reasons the commission had been collecting temperature data from bore holes in every continent. The earth retains a memory of surface changes in temperature and that memory is reflected in tiny changes in temperatures in bore holes. The analysis, 'Late Quaternary temperature changes seen in worldwide continental heat flow measurements', Shaopeng Huang and two others, *Geophysical Research Letters*, 1 August 1997, says that the further down the hole you measure, the further back in time you go. But *Nature* and the IPCC brushed aside that wealth of evidence to seize on the hockey stick, pointing to research showing that the MWP was not evident in some areas, and leaping to the conclusion that the climate pattern really only affected Europe and areas around the North Atlantic. It was made a feature of the panel's 2001 report (the IPCC releases a report every five to six years).

Stephen McIntyre, a Canadian businessman with some science training and a statistical bent, was unconvinced when he saw the graph, particularly as he had just been involved in a business fraud which involved a hockey stick-like graph. He teamed up with Ross McKitrick, a Canadian economist specialising in environmental economics and the pair eventually published a paper, 'Corrections to the Mann et al (1998) proxy data set and northern hemispheric average temperature series' , *Energy and the Environment*, 2003. The also published 'Hockey sticks, principal components and spurious significance', *Geophysical Research Letters*, 2005. These showed that the statistical analysis was wrong and the graph was flawed. The actual flaw has to do with something called a principal components analysis which gave undue weight to tree ring data affected by 2oth century agricultural activity. The pair also found that when the various temperatures proxies used by Mann and others were correctly compiled and analysed the MWP-LIA sequence appeared.

These papers and various complaints provoked quite a storm which eventually resulted in the the US Congressional Committee for Energy and Commerce convening an Hoc committee of eminent statisticians to look closely at the research. Led by respected statistician Edward Wegman of George Mason University, the so-called Wegman report confirmed that the analysis in the original paper contained a basic flaw which invalidated the graph. (*Ad Hoc Committee Report on the 'Hockey Stick' global climate reconstructions*, 2006.) Surprised that other scientists in the same field, paleoclimatology, had not only failed to expose the MBH work but had actively defended it, the committee also spent considerable time and energy examining the relationships between scientists in the field. The analysis identified what amounts to an 'old boy' network which never seriously criticised the research.

Another report issued in 2006 by a committee convened by the US National Research Council of the National Academies chaired by the equally respected Gerard North of Texas A&M University (*Surface Temperature Reconstructions for the Last 2000 years*), does not seem to say anything definite. A discussion on principal components analysis mentions some parts of the work of McIntyre and McKitrick's approvingly but does not say outright whether the hockey stick analysis was flawed or not. Instead, it pats Mann et al on the back for pointing out that warming in the last century had been unprecedented in the last 1,000 years. As that takes us back to the MWP, the observation does not mean anything. However, the report also admits in passing that the MWP and Little Ice Age existed and that some (but not all) records indicate that certain areas may have been warmer in medieval times. In other words, the North committee knew there was something dodgy about the hockey stick but did not want to make waves.

Mann, his co-workers and others have strongly defended the work, roundly criticised aspects of the Wegman report and hailed the North report as a vindication. But after initially also strenuously defending the Hockey stick the IPCC responded to the two reports by apparently pretending that it never existed. The panel did not make any official statement on the issue or any explanation. All any outside observer can say is that the 2001 report heavily featured the hockey stick, but the 2007 report does not mention it at all. In fact, the IPCC and Mann managed to get it into the panel's 2007 report, buried in the section on paleoclimatology, with a tiny bump where the MWP should be. Because of that silence and the fact that the hockey stick's demise attracted very little press coverage, it can still be found in briefing material on climate issues. Activitists were still citing it to me in late 2009.

A great deal more can be said about the politics of this strange episode but it has already been set out in more detail than I could ever manage in Christopher Booker's *The Real Global Warming Disaster*, and Aynsley Kellow's *Science and Public Policy: The Virtuous Corruption of Virtual Environmental Science*, both mentioned in the previous chapter. Booker gives details about how the IPCC managed to sneak the hockey stick into the 2007 report, despite what should have been crippling criticism. If you want a defence of the analysis then read the Wikipedia entry on the issue. That online encyclopaedia can always be relied on for a lengthy, learned justification of any piece of research supporting the climate orthodoxy, no matter how flawed, complete with footnotes.

Several points stand out in this piece of lunacy, which Booker and Kellow also make. Firstly, the Mann et al paper made it through peer review and into *Nature*, one of the two pre-eminent journals in which all scientists want to appear (the other is *Science*), despite contradicting a well-proven climatic pattern and without presenting any new evidence. The paper was simply an analysis of existing temperature records over several centuries, as already done by Huang's team albeit with with just one type of proxy record, borehole temperature readings. The Huang work was arguably far more comprehensive and easier to interpret. About all that can really be said in *Nature's* favour is that the paper was held up for about eight months before being accepted. Further, the MWP-Little Ice Age climate pattern was pushed aside on what amounts to a pretext. There was some evidence at the time that the MWP-LIA sequence was not marked in some areas, but it was marked in many others, and had been further confirmed by the bore hole analysis. However, it can be said that at the time there was little evidence for the climate pattern in the Southern Hemisphere.

Now there is a wealth of evidence that the MWP-Little Ice Age sequence was worldwide, albeit with the MWP ending centuries earlier in some areas and later in others. In *The Little Ice Age* (Basic Books, 2000), by Brian Fagan a former professor archaeology at the university of California, the author describes how a visit to the Franz Josef glacier in New Zealand made him realise that the Little Ice Age was a "truly global phenomenon". He also notes that nine centuries ago, when the MWP was at its height, the glacier was "a pocket of ice on a frozen snow field".

Another piece of research by Thomas N. Huffman of the University of Witwatersrand looked at patterns of millet and sorghum planting as a means of following climate change. Archaeologists can tell an enormous amount about past conditions in an area simply by taking soil samples. They can work out when and where certain crops were planted. Huffman followed the rise and fall of those crops at certain altitudes to conclude that the climate of Southern Africa was characterised by a warm, wet period from 900 to 1300, which he describes as the "Medieval Warm Epoch". ('Archaeological evidence for climate change in the last 2000 years in southern Africa', *Quaternary International*, 1996.) Then there is the work in the Andes published in the online academic journal *Climate of the Past*. After analysing areas occupied by the Incas before the arrival of the Spanish, paleo-biologist Alex Chepstow-Lusty and a team concluded that the area had been warm and productive during the Medieval Warm Period, with the Incas owing their rise to more favourable climate. 'Putting the rise of the Inca Empire within a climatic and land management context', *Climate of the Past*, 22 July 2009).

From four stalagmites in a cave in New Zealand's South Island scientists have reconstructed a temperature record for several thousand years which shows a distinct peak about the same time as the Medieval Warm Period, followed by a cooler period. ('Speleotherm master chronologies: combined Holocene ^{18}O and ^{13}C records from the North

Island of New Zealand and their paleoenvironmental interpretation', P.W Williams et al, *The Holocene 2004*). The ^{18}O and ^{13}C are the isotopes of oxygen and carbon tracked in this analysis. The MWP has been sighted in the Congo ('Culture or climate? The relative influences of past processes on the composition of the lowland Congo rainforest'. Terry M Brncic et al, *Philosophical Transactions of the Royal Society B – biological science* – 2007). That paper also points to a reconstruction of sea surface temperatures off the African coast, from ocean sediment, which also shows a distinct MWP-LIA sequence. A great deal more evidence of the cycles and their world wide spread is cited in the Singer-Avery book.

Both McIntyre and McKitrick have gone on to making a career out of finding statistical faults with various pieces of climate evidence and can now fairly be described as professionals, but for two outsiders – an economist and a businessman – to find such a flaw in a paper produced in Nature is interesting to say the least. They are also known to have had considerable trouble getting Mann to release the code he used for his analysis.

This is further confirmation, if we really need it, that the process of peer review is far from fool proof, and that even the most prestigious scientific journals print research papers that should have been spiked. This problem seems to be more acute in environmental sciences than in other disciplines, and it may not be getting better. As noted above, the NRC report on the hockey stick graph patted Mann et al on the back for various inconsequential matters, but anyone who know the dispute and bothered to read the report would have realised the MBL academics had lost the debate. *Nature's* response was to print a report headlined 'Academy affirms hockey stick graph'(*Nature* June 2006). It would have been better not to say anything at all.

What about the Roman Warm Period? That period may well have been warmer again than the MWP but it is difficult to say as the proxy temperature records are much vaguer that far back and written records sparse, so its generally overlooked. In any case, the argument has changed.

After being forced to concede that the Medieval Warming Period still existed global warmers have taken to arguing that really it was not all that warm and that the climate of the time was unstable. This argument merges with another alleging that the climate signal in the MWP was heterogenous, which is another way of saying it varied from place to place. More variation is a sign of a natural change. In contrast, a more homogeneous climate change – that is, one that varies less from place to place - may be the result of artificial forcing. The 2007 IPCC report presented some evidence that present change is more homogeneous than the MWP warming.

This point is disputed in a paper 'The IPCC on a heterogeneous Medieval Warm Period' in the journal *Climate Change* (2009) by two Swiss researchers Jan Esper of the Oeschger Centre for Climate Change Research and David Frank of the Swiss Federal Research Institute who say there is not enough information from which to draw a conclusion. The Chepstow-Lusty paper cited above acknowledges that the orthodoxy is for medieval climate to be unstable but states that the Incas of medieval times seemed to enjoy a stable climate. Any effort to sort out this argument is well beyond our layperson's tour of the evidence, but we can at least note that there are written records of changes in season covering the period, albeit mostly from Europe, and no one has pointed to evidence of climate instability in those records. In contrast, there is a great deal of evidence that the climate during the Little Ice Age was unstable, with plenty of ups and downs in temperatures causing misery to the Europeans. As the climate science profession (with honourable exceptions) is only fresh from denying that the MWP existed at all, on the strength of flawed research, perhaps we should take further claims with a grain of salt.

In any case, even if modern climate can be shown to be more stable than medieval climate we are still left with a lot of unexplained variations in past climates, which are accounted for by theories other than that carbon dioxide is responsible for everything. These are to be examined in their place.

While on the subject of past climates we will examine various doubtful incidents that came to light while I was writing this book, starting with the tree ring data incident. Academics have for years been publishing data based on rings from trees in the polar Urals in Russia which showed that the modern warming has been sharp indeed. In other words they showed another hockey stick graph. Efforts to get at the data by the McIntyre-McKitrick team were ignored until Keith Briffa of the Climate Research Unit at the University of East Anglica, a greenhouse bastion, and fellow scientist F. H. Schweingruber, published a paper in the *Philosophical Transactions of the Royal Society*. That journal has strict rules on requiring authors to disclose data and ordered Biffra to release the data he used.

McIntyre-McKitrick promptly found that the key part of Briffra's tree ring composite temperature graph, the sharp upturn in temperature in recent times, was based on only a handful of trees. They also found that there were other trees he could have used to give his sequence more depth, but those trees showed no particular trend for modern times. Briffa has since strongly defended himself saying that he chose the trees and analysed the results according to rigorous criteria, and with a view to excluding certain biases. He says his critics did not explain why he should have included the tree rings they referred to.

This dispute is still warm, so to speak (the material above is taken from online exchanges on the issue), and, in any case, has been overshadowed by the juicer so-called 'climategate' incident. In late 2009 computer hackers managed to break into the servers used by the Climate Research Unit of the University of East Anglia in the UK and steal thousands of emails and documents which they posted online. The hackers have not been identified to date, although there is a suspicion that they are Russian in origin. Whoever stole them, the emails between Phil Jones, head of the CRU, which is a bastion of greenhouse science, and other leading global warming scientists, such as Michael E. Mann and American scientist Kevin Trenberth, indicate that they have been using science to play politics. The emails mention using a "trick" to "hide the decline" in a chart on recent global temperature changes. Another from Trenberth, of the US National Centre for Atmospheric Research in Colorado, says that at the moment scientists cannot account for the decline. There is also talk of pressuring editors of science journals to exclude papers that do not agree with the party line on greenhouse warming.

The emails indicate that the scientists in question have been more concerned with the presentation of their data, and on how to explain away the present pause in temperatures, rather than anything that can be described as fraud or falsification. Working out how to overcome problems with a much-loved scientific theory, and excluding papers that do not agree with the orthodoxy, is also hardly new in science as we have seen. With many millions of research dollars at stake some extra angst is understandable. Nor did we really need the emails to show that there was something amiss with the greenhouse orthodoxy or that climate scientists will do a lot to prevent mavericks from marring the 'message' that they want to present. We have already seen this and will see it again in subsequent chapters. However, the incident did highlight the fact that the world is spending billions of dollars and remaking whole industrial sectors on the basis of scientific theory which still has to be defended by such means.

In late 2009, Jones stood down from his position as head of the East Anglia CRU pending an investigation announced by the University. Further developments in this area will have to be left to the media or another book, for there is still a lot of climate theory to cover. As we have seen the earth's climate has varied a lot in the past and carbon dioxide has been blamed for some of those changes, but there are plenty of holes in the theory. There are more holes to come.

Reading isotopes

To track temperatures through hundreds of thousands of years in ice cores and limestone cave formations scientists look at isotope ratios, notably those of oxygen and hydrogen. An atom has a central core surrounded by a cloud of electrons. The bulk of oxygen atoms have eight protons and eight neutrons in that central nucleus, but less than one per cent have eight protons and 10 neutrons. The second type of oxygen atom, an isotope of the first, is chemically exactly the same as the "normal" oxygen atom but has a different atomic weight (the extra neutrons). Water with an ^{18}O atom evaporates less readily and condenses more readily than water molecules composed of ^{16}O. During times of higher global temperatures, sea water tends to be richer in ^{18}O and rain and snow higher. These are obviously infinitesimal differences but can be measured accurately. Once measured the ratio is compared to that of "standard water" which is found in deep off shore ocean water.

In the Vostok ice cores temperatures were tracked through the Deuterium ratio. Hydrogen atoms, the simplest atom of all, had just one proton in its nucleus. The isotope Deuterium, otherwise known as heavy hydrogen, has one proton plus one neutron. Again the chemical properties are the same but a molecule of water with a Deuterium atom has slightly different physical properties. Changes in the ratio of Deuterium water to ordinary water have been found to correlate closely with changes in temperature.

57

3
BAD CALLS

After delving into the last three million years of the earth's climate history we finally arrive at the last 150 years, which have generated all the fuss. Figure six shows the broad sweep of changes in the earth's temperature since 1850, compiled by the Hadley Centre in the UK, a part of the UK Meteorological Office and associated with the Climate Research Unit of the University of East Anglia. Specifically the graph uses the HadCRUTv3 (Hadley CRU Temperatures version three) data series which is by far the most widely used. It was also the most readily accessible until, if you will recall from the previous chapter, the CRU's server was hacked and a host of embarrassing emails leaked. Now it is hard to access basic features on the site.

The line that jumps around a lot involves the actual temperature readings compiled from instruments, which means Hadley researches have collated figures from hundreds of temperature recording stations around the word, including instruments on ships, and adjusted them to produce overall temperature figures for the earth month by month and year by year. They are also available for certain regions.

The Hadley series is, in turn, one of four data series compiled and available from five sites. Three compilations are from networks of ground or sea instruments and two from satellite readings. I have listed these at the end of the chapter. If you grasp the point that there are groups that collect data on temperatures from around the world month by month and year by year, and then compile them into a global temperature figures, you will be ahead of some activists I have encountered who have no idea the figures exist. They have been told the earth is warming and have heard lurid tales of melting ice caps, but they have never realised that there are groups that track global temperatures or thought to ask how the amount of warming has been established.

The smooth line is basically an average of the second – technically a five year moving average – which will give you a much better idea of trends as the unsmoothed data bounces around quite a bit. We will discuss smoothing in a moment as it's a surprisingly important part of the debate. The zero line is simply the average of 30 years of temperatures, from 1961 to 1990, and, like the graphs in the previous chapter, has nothing to do with the zero of freezing water. All figures are expressed as a variation from that baseline. All that need concern us are the changes or trends in the data. Forget the rest.

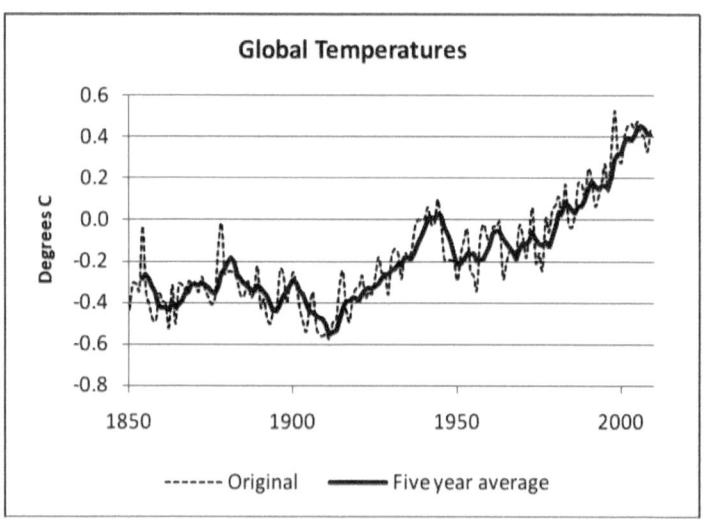

Figure 4: Increase in temperatures since 1850, including the increase between 1975 and 2000. Source: HADCrutv3

Originally I intended to just use the Hadley data rather than get into a lot of boring argument over ground-based instrument networks as opposed to satellite readings. But in addition to the problems with the East Anglia CRU discussed briefly in the previous chapter, considerable doubts have been raised about the operation of the instrument network it uses to collect data.

In the first part of this, in response to freedom of information requests from sceptics, in August 2009 the unit admitted destroying a large part of the raw data collected from the previously mentioned network of instruments. The raw data has to be adjusted for all sorts of reasons, such as when the instrument is moved or the site changes because the area around it has been built up, or if there have been major changes in technology and in methods of collection. Collecting sea surface temperatures decades ago, for example, involved someone from a ship's crew dipping a canvas bucket into the ocean, hoisting it on board and taking a thermometer reading. The subsequent readings had to be adjusted for the heat lost between the water being scooped up, and the temperature taken. Temperature readings in Siberia in the old Soviet Russia used to be influenced by the Soviet habit of subsidising towns in excessively cold regions. Sceptic groups in the US have made it their business to photograph all the ground instruments used in the networks there (the CRU uses data from those instruments as well) and document what changes have occurred around the site since readings began. They report that few sites meet basic standards.

The data was thrown away in the 1980s before Phil Jones was made head of the CRU and when climate was a much less pressing issue, but it is still unfortunate. However, some data seems to have been posted for analysis as in mid-December 2009, after trawling through the data for Russian stations, the Russian Institute of Economic Analysis accused the CRU of cherry-picking the data by only using those stations that showed a warming trend. An English language summary of the Institute's paper is posted on the site of the Cato Institute in the US. Those who wish to brush aside such reports will immediately scream that the Cato Institute is very right wing and that the Russians have remained notorious sceptics throughout the debate, and would be right on both counts. The Russian Academy of Sciences has notably refused to endorse the global warming orthodoxy. But before we get bogged down in the politics there is more to come. In late January 2010 two American broadcast meteorologists and prominent sceptics

Joseph D'Aleo and Anthony Watts released a lengthy paper saying that the Global Historical Climate Network, which several centres rely on for much of their land data, has lost about three-quarters of the 6,000 stations that use to feed data into it.

The weather station network has been rationalised since the 1980s because of the reliance on satellites but it is the more remote stations in cooler locations such as Northern Canada that have gone, and those near urban centres which are generally warmer have stayed. This would not matter so much if the results from the network had been adjusted, but D'Aleo and Watts say that there has not been any adjustment. (*Surface Temperature Records: Policy Driven Deception?*, Science and Public Policy Institute, 2010.) Watts' site, Watts Up With That is one of the sceptical landmarks in the global warming debate, and he has been one of the leading lights in the push to inspect all the data recording sites across America.

On top of all that, CRU's Jones has become involved in a bizarre academic dispute over where exactly are, or were, 42 weather monitoring stations in remote parts of rural China. (*The Guardian*, 1 February 2010). The dispute involves a paper written by Jones and others back in 1990 published in *Nature* which showed that the "heat island effect" was unimportant. Temperature recording sites in areas that get built up will show a warming effect simply because buildings have been constructed and asphalt laid close to it, quite independent of any warming in the climate. Jones has since said that his university has investigated the matter and exonerated him of all allegations.

We are not going to wade through the issues raised above. These sorts of discussions can quickly become boring and we are big picture people. So instead of just using the Hadley data as I planned to do I have also included satellite data compiled by the University of Alabama in Huntsville (UAH – see notes at the end of the chapter) from a NASA satellite. This is not ground temperature data, but readings of the lower atmosphere. However, the data has been audited and tells a similar if less dramatic story as far back as it goes, which is 1978 when satellites were first used for this sort of work. You will recall carbon dioxide is supposed to be warming the earth, so if there is additional warming it should be in the atmosphere. The satellite data is below.

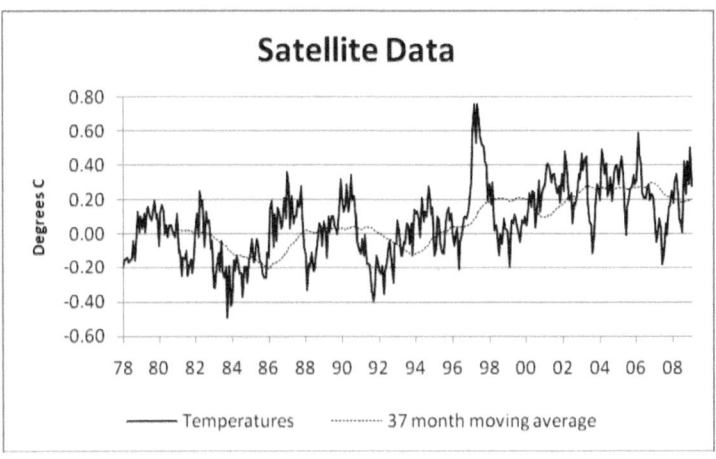

Figure 5: Thirty years of data from satellite readings of the atmosphere show far less warming than instruments on the ground. Data is from the University of Alabama in Huntsville.

The smoother dashed line in the graphic is just the three year (plus one month to keep the total odd) average calculated for each month, and again is just there to show the trends. An important point to note is that dips and spikes inside those 37 months are smoothed over so that you are left with a small trough or a bump. Increasing the smoothing period to five, seven or even 11 years will iron out much bigger spikes and falls. A point to which we will return.

From the first graph you can see that the earth has warmed over the past 160 years which no one disputes. By 1940 temperatures had increased by about 0.4 degrees or four tenths of a degree, but then fell away with that decline causing the previous mentioned talk about a looming ice age. Note that at the time the temperature dip was previously believed to be much smoother. The hockey stick graph we saw in the previous chapter, using instrument records of the late 1990s, shows a distinct double curve for the same period. The figures seem to have since been changed or corrected, but perhaps corrected fairly. Note that the graph results from the 19th century in particular required scientific forensic work on old meteorological records but, despite shortcomings, the quality of the data from the old instrument record is a world above that of data derived from fossil records. Its resolution is notably higher.

From the mid-1970s to the turn of the century a major warming spurt added about 0.6 degrees or six tenths of a degree to the total, making the overall change between 1840 and the end of the first decade of the 21st century, 0.8 degrees. The overall result is eight tenths of a degree in almost 170 years, albeit with the bulk of the warming occurring in just 25 years. This coincides with a lot of emissions and much of the global warming case rests on that 25 years of warming. The satellite data in figure seven shows that there has been much less warming in that same period, perhaps four tenths of a degree. The change is much less noticeable on the graph because it spans just 30 years whereas the upper graph spans 160 years. Both graphs agree that since the end of 20th century, there has been very little warming and even mild cooling in the past few years.

As you can see from both graphs 1998 was a monster year for warming, with that spike attributed to a climate cycle called El Niño. The surge in the monthly temperature records at the end of the satellite data (which is also reflected in the monthly data from the Hadley Centre) is attributed to another El Niño event.

An Australian Bureau of Meteorology information release says that the term El Niño (Spanish for 'the boy') refers to a situation when sea surface temperatures in the central to eastern Pacific Ocean are significantly warmer than normal. This occurs every three to eight years. Under normal conditions, a stream of cold water wells up from the deep ocean along the Peruvian coast and so the eastern and central Pacific is eight to 10 degrees colder than the western Pacific, around the top of Australia. That difference in sea temperatures, in turn, drives an atmospheric circulation known as the Walker circulation. Air rises over the warm part of the Pacific with one result being towering cumulonimbus clouds and rain, some of which falls on Australia. The air then moves east to sink over the colder part of the Pacific. In El Niño years the cold water flow weakens and may even vanish completely so that the central and eastern Pacific warms up a lot. That change greatly alters the circulation patterns, meaning that Australia gets less rain but, more importantly for our discussion, a large area of the ocean becomes warmer so overall the world becomes warmer. There may also be knock-on effects depending on ocean-atmosphere cycles in other parts of the world, although those effects are controversial and only vaguely understood. Whatever the reason, in 1997-1998 the El Niño event caused a major spike in temperatures.

The converse, La Niña (Spanish for 'the girl'), involves a much stronger upwelling of cold water and so a colder eastern and central Pacific, and stronger resulting air circulation. That mostly means more rain for Australia. A La Niña event resulted in the sharp fall in month on month temperatures you can see in the satellite data (and also in the Hadley monthly data) in late 2007 and early 2008. Another, prolonged La Niña period in 1998 through to 2000 probably caused the dip in temperatures visible on the graph for those years.

Both climate cycles can be tracked by an index called the Southern Oscillation Index (SOI) which, despite its fancy name is just the difference between air pressure readings in Tahiti and Darwin. The air pressure reading in Darwin is subtracted from the readings in Tahiti. If the result is zero then everything is normal but when El Niño events occur, the air pressure in Tahiti is lower than that of Darwin, meaning that the normal circulation patterns have changed, and the SOI is said to go negative.

The La Niña and El Niño events are by no means all there is to Australian climate, as it is possible to have floods during a La Niña and dry years during an El Niño, but they are important influences. There is a distinct correlation between the SOI and Australian wheat harvests. By watching ocean temperatures through a network of moored buoys carrying instruments, Australian BoM scientists can greatly improve their seasonal forecasts. Although they cannot yet forecast when one of these cycles will start, they know when it does start. All this is good but we are still left with lots of questions. Why did the El Niño in 1997-98 have such a big effect and the 2009-10 event have so far comparatively little? Or are these cycles part of the effects of larger cycles rather than the causes? Bear in mind that the cold water up welling off the Peruvian coast has to come from somewhere.

There are lots of other ocean-atmosphere climate cycles – in fact a whole zoo of them – which are now being investigated. The El Niño Southern Oscillation to give the El Niño-La Niña cycle its formal name is one the shortest. The Pacific Decadal Oscillation, the Arctic Oscillation and the Atlantic Meridional Oscillation have cycle times of decades, and potentially have much bigger effects, which we will discuss later in the book. For now we will note that these cycles exist and move onto the big picture climate forecasts that have caused all the fuss.

Stepping back from the graphs for a moment the total increase of 0.8 degrees over 160 years is not impressive, particularly given the IPCC forecasts for temperature increases to come. The 0.6 degrees in 25 years up to 2000 or so is more of a concern but bear in mind that we are looking at this climate data in high resolution and at an unusual time, when the earth is recovering from a cold period. There is nothing in past records to indicate the change is unusual and, in any case, nothing much has happened for the past decade or so. The satellite results show two-thirds of the increase of the instrument data so if we rely on that data then hardly anything has happened, and there are scientists who say that only the satellite results can be trusted.

Nonetheless other scientists have been inspired by the apparent big increase in temperatures in the quarter of century up to 2000 into building computer models using complex equations to simulate the earth's climate, and to churn out alarming temperature projections which they insist must be right. Just look at how much temperatures increased in just 25 years. It is a sign.

One of the first cabs off the rank in this respect was James Hansen, director of NASA's Goddard Institute of Space Sciences, who built the NASA-GISS model 1988 (there are now several other groups that model climate). In that same year he testified before a US

Senate Committee chaired by Senator Tim Wirth of Colorado. Wirth was determined to push the line that greenhouse gases were warming the earth and invited Hansen who was known to strongly favour the concept to give evidence to the committee. In a piece of theatre the senator's staff scheduled the hearing on what they hoped would be a very hot day in summer, which it was, and then opened all the windows in the hearing room. The director's presentation, which had been given before, was also brightened up and simplified for general audiences The extra presentation punch resulted in wide publicity for the message that the earth was warming up and that industrial activity was to blame. Of course the earth was warming at the time. The real issue was whether the temperature increase was in any way unnatural, or any part of it had been forced by human activities. Twenty years later, many of the consumers of global warming messages have still to grasp that point. The assumption is that because the earth has warmed up a notch human activity must be to blame in some way.

The congressional hearing and the resulting publicity helped push the United Nations Meteorological Organisation and the UN Environmental Programme into setting up an intergovernmental panel which eventually became the Intergovernmental Panel on Climate Change. We will skip over most of the irksome political details, which are all dealt with elsewhere and look at the IPCC's output. These are mainly a series of what the panel calls assessment reports all warning that human activity, namely industrial emissions, is artificially warming the climate. The first of these was issued in 1990, with others following in 1995, 2001 and 2007, plus some supplements.

The first report is no long available to download from the IPCC site but the basic forecast of that 20 year old report can be found by scouring the internet, and a glance shows why the panel is no longer keen on having it around. The report says: "Based on current models, we predict: under [BAU – business as usual] an increase of global mean temperature during the [21st] century of about 0.3 $^\circ$C per decade (with an uncertainty range of 0.2 to 0.5 $^\circ$C per decade)."

That seems to read that the panel is expecting an increase of a minimum of 0.2 degrees (two tenths of a degree) per decade, so by that forecast temperatures should have increased by at least four tenths of a degree (0.2 degrees per decade for two decades) since 1990, but clearly it has not done so, not even by the Hadley graph. It's not right for either of the two decades since the report. If we use the satellite data the difference between forecast and result is even worse. One activist response to this I have encountered is that only a trained climatologist can tell us whether the forecast is a success or a failure. Nonsense! A forecast is a forecast; the figure given is easily understandable and the actual results are readily available. As we saw in previous chapters there is no guarantee a forecast made with the best scientific expertise available will be right, so we should have no hesitation in pointing at this forecast and saying "it's wrong".

Another response to this rarely mentioned failure is that the IPCC forecasting and reporting process has become increasingly more sophisticated, and the scientific understanding of climate has increased. Activists also know that industrial emissions must be to blame for something. So the panel has kept churning out reports. The IPCC does not conduct any research itself or monitor climate. Instead it relies mainly on voluntary work by scientists around the world to compile and analyse scientific work done by themselves and others, and run the all important computer projections. Each report has become a vast undertaking of reviewing and collating projections, plus reporting on the various physical effects expected to result from the forecast changes.

Although, as noted, the IPCC reports have been forecasting dire temperature increases from the beginning, most of the fuss has been over the 2001 or Third Assessment Report (TAR), and the 2007 or Fourth Assessment Report (FAR). The release of the 2007 report (I will always use the dates, rather than the initials in referring to these reports) came after the 2006 documentary *An Inconvenient Truth* about the campaign by the former vice president of the United States Al Gore to raise awareness over the supposed human warming of climate. Then there was the report from the UK Government called the Stern Review on the Economics of Climate Change also produced in 2006. Stern will be dealt with in another chapter but all these reports created a perfect storm of concern over climate change.

In that storm, the many concerned citizens and a lot of scientists seem to have overlooked – or perhaps not fully realised - that they were dealing with a bunch of forecasts from unproven forecasting systems. That is they are unproven in the conventional sense of having made forecasts that had been found to be tolerably right in the course of time. In fact, as we have seen, the earlier forecasts were wrong. Instead, scientists who had built systems modelling the vast interplay of winds, clouds, sunlight, ocean currents, ice, land and mountain ranges, and clouds that govern climate, have managed to work them so that they produced results which fit for the past few decades or so of temperature results. This is called hind casting. The same sort of technique is used for stock market trading models, incidentally, and called back testing. But once you have finished back testing or hind casting the idea is to use the resulting models to see if they can make money or, in the case of the climate models, to see if they have anything useful to say about unknown data. As we saw in the chapter on 'Forecasting Follies' a successful hind casting exercise means nothing. We have to look forward.

The 1990 report forecasts were not worth the trouble, and the 1995 report has been taken away so to evaluate forecasts made by these more sophisticated computer models, let's look at the 2001 report, which is really the only game in town. The 2007 report is just three years ago, at the time of writing, and that is not really long enough for a proper comparison. In addition, there is an attempt in the scientific literature to compare the 2001 report with what has happened since.

The 2001 forecasts start from 1990. In figure six I have put in the upper and lower limits of the forecasts given in the 2001 report (available from the IPCC web site www.IPCC.ch) through to 2010, and compared them with the Hadley figures. I started the IPCC forecasts and the five year average off from the same point, which requires a simple adjustment.

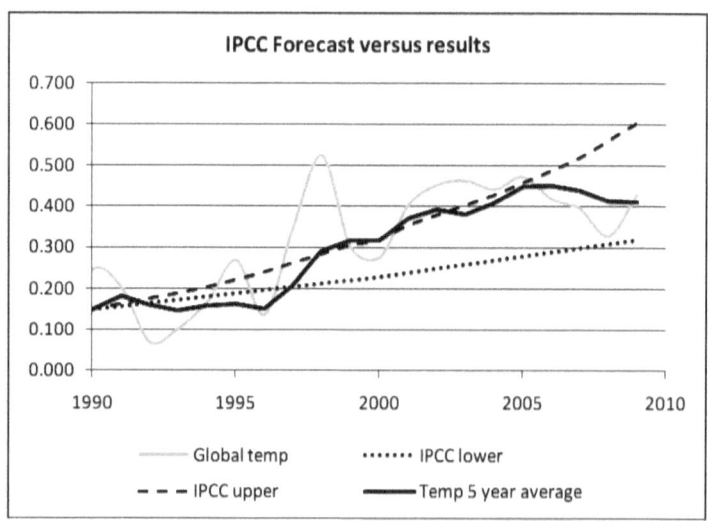

Figure 6: Comparing IPCC 2001 forecasts with results. Note that in 2001 the average was above the top line of the forecast.

As you can see from the above straightforward (unkind people might say unsophisticated) analysis up until four of five years or so ago it was possible for global warmers to claim that measured temperatures were actually at the top of the panel's 2001 forecasts. This point was the subject of a paper 'Recent Climate Observations Compared to Predictions' (*Science*, 4 May 2007), with the conclusion helped along by the use of a much longer smoothing interval of 11 years and the temperature series compiled by the Goddard Institute of Space Studies (see note at end). The paper, which lists nine authors led by Stefan Rahmstorf, a Professor of Physics of the Oceans at Potsdam University in Germany and including the GISS director James Hansen, has occasionally been cited as evidence that the earth is warming faster than expected.

Unfortunately for the warmers it is not that simple. Although I have included the figures from 1990, in fairness to Rahmstorf's approach, the only true test of any forecast is how well it did in with figures that were unknown at the time it was made. The report was issued in 2001 so its authors knew the 2000 figures. A closer look at the graph shows that five year mean was at the top of the forecast range at the time it was made, so it is hardly surprising that it took a little time to diverge from the upper line. In fact, the bulk of the warming took place before the forecast was made. Also, as we can see by simply inspecting the graph, there are indications that instead of temperatures accelerating to some tipping point they are coming off a peak. If we had used the satellite data instead of Hadley the forecast would be dead in the water.

For those looking forward to a temperature Armageddon this is not an impressive result particularly given the way the forecasts were started at the top of the range, but the difference between a good result and a bad one is small. The comparison can also be improved, or made worse, by changing the graphic. An informally produced *Cherry Pickers Guide To Temperature Trends (down, flat and even up)* by Chris Knappenberger on Master Resource, an energy blog (http://masterresource.org), points out that it is possible to make cases for current temperatures being both almost in line with 2001 forecasts, or well below forecasts, by picking the temperature series and the smoothing interval. In my unsophisticated way I have used a five year smoothing period, but an 11 year smoothing period can be justified and makes the comparisons look much better. However, most readers should be able to read graphs for themselves, without having to refer to highly skilled climatologists for guidance. What does it look like global temperatures are doing in the first two graphs? As temperatures are, in any case, now below where they were in 2001 and 2007 a common sense view is that the forecasts are not successful.

Our common sense approach also deals roughly with the longer term forecasts. The 2001 IPCC forecasts were for temperatures to increase by anywhere between 1.5 degrees to almost 6 degrees over a century, but the more likely result was considered to be between 2 and about 4.5 degrees. To reach the top of the broader range forecasts temperatures would have to increase by 0.6 or six tenths of a degree each decade, although the increases would be less at the beginning of the period and more later on. In the noticeable temperature surge between 1975 and around the end of the century, and again using Hadley, the rate of increase may have briefly touched 0.3 degrees a decade. As the total increase was about 0.6 degrees, over 25 years that works out to an average of 0.24 degrees per decade. Again you can look for yourself on the graphs.

In other words the IPCC issued its forecast expecting that temperature increases evident in the previous 25 years would at least be the same, or accelerate markedly, because more CO_2 was pouring into the atmosphere from smoke stacks. Instead, as with the forecasts of the American criminologists and solar scientists, the ink on the IPCC report was barely dry before the actual system started doing something different. The claim about hind casting also presents a problem. As we have seen when the 2001 report was issued, the scientists involved with it were convinced that there had been no Medieval Warm Period and the hockey stick was the prevailing wisdom. So the temperature graphs using the models back past 1900 in the 2001 show a flat line. As noted, the 2007 report put in a token bump in the relevant temperature graphs for the MWP, but gets around the problem of not being able to hind cast anything past 1900 by simply cutting off the results of their computer models at that date.

Despite changes in the theory the 2007 IPCC report said much the same things as the 2001 report but cut the top range increase to about four degrees for the century, and the bottom range to a mere 0.6 degrees. As you will note both forecasts cover a wide range of possibilities; all the way from a mild warming which we would probably not notice to very substantial increases indeed. Given such a wide range of forecasts, surely the final result must be somewhere in that range, particularly as industrial activity is pumping a lot of carbon dioxide into the air? We will return to that point, but for moment we only need take heed of the fact that warming was supposed to have accelerated due to increases in CO_2 and it hasn't.

None of this adds up to much encouragement for the warming side, although it is still possible for scientists to say that present temperatures are within the forecast range. To maintain that line for much longer, however, they need temperatures to turn up. In the last few months before this was written the El Niño effect pushed up global temperatures somewhat, albeit not as much as the global warmers had hoped. Once that effect works its way through the system – perhaps by the middle of next year, but no one really knows – temperatures may continue to decline (a point we will discuss in the next chapter). Such a decline will generate considerably more of the scepticism that global warmers have found so annoying. The general drift of scientists away from the global warming cause, thanks to the decade long hiatus in temperatures, will become a flood.

Global warming activists and deeply committed scientists do not like to hear this reasoning, and some of them have not heard it at all. One scientist I spoke to a couple of years ago, before the cooling trend became more evident refused to believe that there had been any cooling at all. When shown a Hadley Centre graph illustrated the cooling (in the southern hemisphere at that time) he dismissed it, pointing with some force to the undoubted temperature increases up to the turn of the century. Activists have more recently accused me of truncating data sets, pointing out that the temperatures in 2009 are still well above those of 1975. Therefore temperatures have still increased, overall. That is right, but misses the point. You cannot justify a forecast by pointing at what happened before the forecast. We already know what happened. The point is do we know enough about the climate system to make successful forecasts? To date the answer is no.

The UK Meteorological Office, a greenhouse bastion, has not helped matters for the global warming side by producing a series of inept seasonal forecasts. One UK media report says that the Met failed to predict wet summers for the past three years; and that its annual global forecast predictions were wrong for nine of the last 10 years (*'Met office deserves to be shown the door', The Independent*, 19 January 2010). In 2009, for example, the office forecast a "barbecue summer", only for the actual summer to be milder and much wetter than previous summers. But the real blunder was a press release issued in December 2009 once it was evident that the El Niño effect had emerged, saying that that "a combination of man-made global warming and a moderate warming of the tropical Pacific Ocean, a phenomena known as El Niño, means that it is very likely that 2010 will be a warmer year globally than 2009". In fact it expected the year to be warmer than 1998, the hottest on the instrument record. ('Climate could warm to record levels in 2010', Met Office, 10 December 2009) The Met Office went on to say that a record warm year is "not a certainty", as the El Niño cycle may give out early, and was referring to 2010 as a whole rather than the first few days of the year in the UK. But the release's timing was most unfortunate, as it had barely been issued before Britain was literally buried in snow, and exceptionally cold conditions had grounded airline traffic and stopped public transportation systems.

This proved too much even for the BBC which has faithfully reported the greenhouse line. In an interview gleefully shared around the growing network of sceptic newsletters and blogs Met office head, John Hirst, was grilled by a BBC presenter over these forecasts (broadcast 7 January 2010). The interview concentrated on Hirst being awarded a bonus for the previous financial year and confused the office's short term weather forecasts, which have nothing to do with global warming, with its seasonal forecasts. It also did not get any answers. The Met office is hardly alone in making poor seasonal forecasts, incidentally. In 2007 the New Zealand Climate Science Coalition, which is definitely not pro-warming, issued an analysis of seasonal climate predictions by the country's National Institute of Water and Atmospheric Research, which found that the overall accuracy of the predictions was just 48 per cent. ('World climate predictors right only half the time', 8 June 2007). There is no reason to think anyone has a better success rate.

Now we come to the real question. The very basic analysis of climate forecasts to date, by treating them as forecasts without going into details, does not show them in a good light at all. So why are we bothering with them at all? At most there may be a cause for concern and further investigation, but they cannot be taken seriously. They cannot be used to justify billions of dollars worth of expenditure, or the complete remaking of industrial sectors and major efforts to cut back on everyone else's living standards. One response to that bleak assessment is to claim that it is all inevitable and it's all in the physics. We are spewing out vast quantities of industrial gases, carbon dioxide levels are known to be quite high and have increased of late and temperatures have also increased. Carbon dioxide is a greenhouse gas, so that proves it. One scientist, in a lengthy exchange of emails with me on the subject in 2009, claimed that "all the equations are well known".

Even the scientists seem to think that more CO_2 must equal more warming so it's all inevitable. As we shall see it's not that simple.

Centres that track global temperatures

There are five recognised centres which publish temperature figures for both the globe and for different regions. Three of these collate and analyse temperatures from a spread of ground based instruments, mainly those run by the Global Historical Climatology Network. The GHCN is, in turn, run jointly by the National Climactic Data Centre at Arizona State University and the Carbon Dioxide Information Analysis Centre. Depending on the centre this data is combined with measurements by ships, which are now automated. Satellite data has been used at times.

Hadley - Generally regarded as the most authoritative of the centres collecting data from ground instruments, it is a part of the UK Meteorological Office and closely associated with the University of East Anglia in the UK, which are both IPCC bastions. As noted some questions have been raised about GHCN and Hadley's approach to the data, so the chapter uses both Hadley and the UAH satellite data.

GISS - Goddard Institute of Space Studies. A part of NASA. The director of this instrument centre is arch greenhouse spruiker James Hansen. Greenhouse proponents always quote this centre's temperatures for the annual results from this site are higher than the others. Also, the GISS records show the highest recorded recent temperature occurring in 2005, while all the other sites show it to be

1998.

NOAA - National Oceanographic and Atmospheric Administration. Owned by the US Department of Commerce. It uses ground instrument data.

UAH – The University of Alabama in Huntsville is known for its science and aerospace programs. The satellite centre is run through the Earth System Science Centre. Director Dr John Christie is sometimes cited as a greenhouse agnostic, although he has contributed to several IPCC reports. Temperature increases recorded since 1975 when satellites started looking at the atmosphere, by measuring microwave radiation in certain bands, are not as dramatic as those shown by the ground based instruments.

RSS - Remote Satellite Services. As the name suggests it is another site that analyses data from satellites. Its results are similar to those of UAH.

4
MODEL MAYHEM

Most consumers of climate forecasts think of CO_2 as akin to a woolly blanket and that adding more makes the blanket thicker, but it is not like that at all. The physics dictates that if we treat the atmosphere as a pile of air, as if it is in a greenhouse, then once CO_2 concentrations are past a certain point the warming effect from the gas taper off. Doubling CO_2 concentrations will not double the amount of warming.

When the earth heats up a certain amount it gives off heat in the form of infra-red radiation, and CO_2 absorbs that infra-red radiation on certain band widths. But the gas does not retain that energy. Instead it is re-emitted, with some of the radiation going towards space and some towards earth. There are some more complications, with scientists talking about the earth's radiative balance, but the bottom line is that only small concentrations of the gas in the atmosphere are required to block most of the radiation. Adding more does not change much. This point isn't in dispute, it is just never mentioned. If everything else remains the same and the atmosphere is treated as a pile of air, then if the concentration of CO_2 in the atmosphere increases from the accepted pre-industrial level of 280 parts per million (ppm) to the present 387 ppm (which works out to 0.0387 per cent), the result should be an increase of 0.7 degrees. The actual result was 0.8 degrees but seemed to vary with no real relation to increasing

CO_2 concentrations. If existing concentrations are doubled we can expect a temperature increase of 1.1 to 1.7 degrees, with the most common figure being 1.1 degrees.

As even scientists unfamiliar with the field have been incredulous when confronted with this well known effect of CO_2 we will check a few sources. The high end of the figures above is estimated off a fully worked out graph for temperature changes versus carbon dioxide concentrations in the previously mentioned classic *Climate History and Modern World*, by H. H. Lamb. Activists can refer to 'The ice-core record: climate sensitivity and future greenhouse warming' by a group of French scientists led by C. Lorius (*Nature*, 13 September 1990), which gives the increase resulting from doubling present concentrations as 1.1 degrees. Sceptics can check *The Climate Caper* by Garth W. Paltridge which also gives a figure of 1.1 degrees. Paltridge also says the change would take centuries.

To illustrate the point further, and again estimating from the *Climate History* graphic, the first 400 parts per million of CO_2 warms the system by a little less than seven degrees. Adding another 400 parts per million, to boost concentrations to 800ppm, or 0.08 per cent of air, increases temperatures by perhaps another 2.0 degrees. Again that is if the atmosphere is treated as a pile of air. *A Primer on CO_2 and Climate* by Howard C. Hayden does not give a figure but points to the tapering off effect of increasing CO_2 concentrations. (Yes, Hayden has some connections with the energy industry. This is irrelevant. See the comment in the chapter *Show Me The Money!*) In his book *Climate Confusion*, the previously mentioned Roy W. Spencer, formerly a senior scientist for climate studies at NASA, puts the temperature increase for a doubling of CO_2 concentrations in the air at one degree Fahrenheit or a touch over half a degree Celsius. Spencer does not give any details in the book, which is a useful guide to the issues for the layman, about just what that estimate applies to, but the figure is in the ballpark for concentrations moving from 387 to 560 ppms, or double pre-industrial levels of the gas.

That's not quite the end of it. Sceptics occasionally allege that measurements of the infra-red band on which CO_2 absorbs radiation show that it nearly saturated, meaning that the gas cannot cause very much more warming. On the global warming side, a book available online *The Discovery of Global Warming*, by Spencer Weart, which has useful points to make about the history of the physics of all this, seems to imply that extra CO_2 in the upper atmosphere causes additional warming.

As we can see it is generally accepted that the warming directly caused by CO_2 is limited. So how do the model forecasts get from under one degree to four or so? In part by assuming that the warming effect of the CO_2 is amplified, just as the change in solar radiation from the cycles discussed in previous chapters is supposed to be amplified to make changes far in excess of the direct effects. Here we come to the much discussed issue of climate sensitivity to increases in CO_2. Climate models are far too complex to be easily dissected or critiqued even by other climate modellers, but it is generally accepted that the IPCC climate models amplify the CO_2 warming by inserting a feedback effect from water vapour. The warming from CO_2 is expected to trigger changes in the amount of water vapour (humidity, clouds etc.) in the atmosphere and that extra water vapour causes more warming, and that warming leads to more water vapour and so on. Water vapour is the real greenhouse gas, at least in the computer models. Additional water vapour should mean more clouds, but the models also assume that those additional clouds will not cause any cooling.

Chapter eight in the IPCC's 2007 report states that "vapour feedback is the most important feedback enhancing climate sensitivity" and discusses the issues of clouds. That chapter also defends the models at some length saying that now more of them do not require "flux adjustments" (fudging) to maintain a stable control climate and are good at modelling the present climate. They also say that they can be made to simulate past climate and temperature changes – a point to which we will return.

Whether you agree with this defence or not, these models are known to be full of assumptions. Some of these assumptions are due to the natural system being so big that they have to be simplified just so that they can be run in a reasonable time, even on a super computer. Other assumptions are there because little is known about that aspect of the physical system, despite constant assertions that the physics is settled and certain. Of these assumptions arguably the most important, and certainly the most discussed, concerns water vapour. Despite all the screaming and shouting over the IPCC forecasts I believe that very few of the panel's supporters, even among the scientists, are aware of the assumption's existence or of the arguments over it. Nor would they realise that very little is known about the details of precipitation systems (clouds and rain), and their interaction with atmospheric humidity. Will there be any extra water vapour to begin with and why is it there? Will the extra water vapour simply fall out as rain? The extra clouds implied by the additional water vapour should cause cooling, or are we missing something? The best explanation of this assumption is in a paper written for the US Environmental Protection Agency by Alan Carlin, a senior operations analyst, and John Davidson in the US Environmental Protection Agency's National Centre for Environmental Economics. Carlin is an economist with a degree in physics and Davidson holds a doctorate in physics.

Written in haste in 2009 to meet an internal EPA deadline, the paper comments on efforts by the EPA to declare that a series of greenhouse gases, including carbon dioxide, be declared pollutants. (That declaration has since been made.) In the paper, which was blocked by EPA management but circulated informally, Carlin notes that much of the warming that the climate models forecast from increases in carbon dioxide can be traced back to a National Academy of Science's (NAS) 1979 study often referred to as *The Charney Report*. This report made similar forecasts to the 2007 IPCC report by assuming that the Relative Humidity (RH), that is the proportion of water vapour in the atmosphere, would remain the same when the atmosphere warms up. But the warmer the air the more water vapour it can hold, and so more water vapour is required to keep the same RH.

Water vapour is the major greenhouse gas, so the models show considerable warming. Carlin says that some of the early models such as the NASA-GISS model compiled by James Hansen in 1988 went even further by requiring RH to increase for certain sections of the atmosphere. "No wonder Hansen got such high values of global warming for doubling CO_2 (in those early models). This logically followed from his extremely high and unrealistic water vapour assumptions," Carlin says.

The water vapour assumption is, in turn, just one of many variables that can be "tweaked" or adjusted to fit whatever the model is supposed to do, because so little is known about the variable. In the previously mentioned *The Climate Caper*, atmospheric scientist Garth Paltridge calls these variables "tuneable parameters". Paltridge, a former director of the Institute of Antarctic and Southern Ocean Studies, says that the modellers usually don't bother to mention these parameters but they are one of the reasons that they produce such a wide range of forecasts. "The complexity of the models is so great that is extremely difficult even for climate modellers to establish exactly why one model should give a vastly different answer to another – let alone establish which is more likely to be correct."

Besides being so adaptable that they can give any result required including a fit to any past climate, climate models are also known to be extremely sensitive to initial starting conditions (they all have to start somewhere). It has never been a matter of plugging a few numbers into an accepted computer model and then proudly displaying the results. Humidity is just one of the assumptions under discussion – others include the behaviour of clouds, rainfall, ocean circulation and ocean climate cycles - albeit a key one.

A paper presented to the 2nd annual conference on Climate Change in 2009 (sponsored by the conservative Heartland Institute in the US and so decidedly anti-IPCC) William M. Gray, professor emeritus, Department of Atmospheric Science, Colorado State University and previously mentioned as a leading sceptic, points out that observed trends on water vapour in the atmosphere do not agree with the forecasts.

In the paper Gray says that models assume that a slightly stronger hydrological cycle (water-evaporation-clouds-rain) due to extra CO_2

will cause additional upper-level atmospheric water vapour and cloudiness. "Such vapour-cloudiness increases are assumed to allow the small initial warming due to increased CO_2 to be unrealistically multiplied 2-4 or more times... Observations of the upper troposphere water vapour (the troposphere is the atmosphere up to a high altitude) over the past 3-4 decades from the National Centres for Environmental Prediction/National Centre for Atmospheric Research (NCEP/NCAR) reanalysis data, and the International Satellite Cloud Climatology Project (ISCCP) data show that upper troposphere water vapour appears to undergo a small decrease, while outgoing long wave radiation (infra-red) undergoes a small increase. This is opposite to what has been programmed into the GCMs (the climate models)."

Grey goes on to make other points, such as the models missing out much of the fine details of modelling clouds because the "grids" used in the computer simulation are too large (that is, a very large cube of atmosphere is treated as one data point for the purposes of the simulation), but reserves most of his ire for the water vapour issue. "Not only have Hansen's extreme and unrealistically high values of upper troposphere moisture not been challenged they were instead closely emulated by most of the other prominent GCM groups," he says. "The fact that most of the (assumed independent) GCMs produced similar warming results were used as verification of each model's results. But this was untrue. All the modellers were wrong in the same direction and in the same way."

The point about water vapour content has been raised many times before but generally countered with claims that it is not in the refereed literature, or that other papers in the literature question the trend. As one piece of evidence modellers can point to a paper by Amato T. Evan and two others 'Arguments against a physical long-term trend in global ISCCP cloud amounts' (*Geophysical Research Letters*, 17 February, 2007). The paper argues that a long term decline in "cloud amounts" (where clouds are an indication of humidity) found by ISCCP is simply a result of the way the satellite observes the atmosphere. The results are an "artefact" of the instruments and not a real trend, and therefore not of a concern.

Sceptics acknowledge that the satellite data can be tricky, but there is other evidence from instruments tied to balloons and sent high into the atmosphere that water vapour in the atmosphere above an altitude of three kilometres is decreasing. In *The Climate Caper*, Garth Paltridge says that in 2008 he and two others analysed the previously published data from many years of such observations to demonstrate that there was a decrease in high altitude water vapour, and sent it off to a climate journal (he does not name the journal). As part of the peer review process the paper was sent to two referees, and one of those sent back an "hysterical" attack which included the statement that the paper could be used to criticise the IPCC temperature forecasts. The paper did not get published. Paltridge also says he later told this story to a small workshop on atmospheric humidity held at Lamont Doherty Earth Observatory in New Jersey, where the attendees split over whether the trend should be reported. Those saying it should not claimed the data was too shaky to be made publicly available, because it could be used to question the IPCC forecasts.

For its part the IPCC is fairly optimistic about water vapour. Chapter three of the 2007 report does not have much definite to say about water vapour in the lower troposphere (the part of the atmosphere below the stratosphere). In the upper troposphere it says, "to summarise, the available data do not indicate a detectable trend in upper-tropospheric relative humidity. However, there is now evidence for global increases is upper-tropospheric specific humidity (there is more water vapour up there), which is consistent with observed increases in tropospheric temperatures and the absence of any changes in relative humidity. It also says that there is no consensus over the change in cloudiness on a time scale of decades.

Whatever the results of those arguments we are still left with the problem of some evidence against humidity increasing, as opposed to not very convincing or definite evidence for increases in relative humidity throughout the atmosphere, and nothing much at all confirmed for clouds. As we have previously seen forecasts involving natural systems are fragile things in which Murphy's Law (anything that can go wrong...) operates very strongly. They can be wrecked by almost anything. In his *Climate Confusion* Roy Spencer says that even a slight change in the size of raindrops modelled can dramatically change the amount of water vapour in the atmosphere.

With forecasts so fragile and so many billions of dollars at stake it is not good enough to say that there is nothing in the literature or that two pieces of evidence cancel each other out, so a key assumption has not really been contradicted. Nor is it really good enough to say that the models agree with one another when they make the same assumptions. The IPCC modellers should get busy and properly confirm key assumptions and, while they are at it, work out some way to satisfactorily confirm their models with forecasts that are not equivocal or arguable, or have to be fudged before being presented. Further, those forecasts should be presented to a truly independent panel for confirmation. Then they can issue temperature projections for decades into the future, although those forecasts will probably still be wrong. A useful rule to adopt concerning forecasts in any field is that they are wrong, until proven right – and even then they might be right for the wrong reasons.

A further discussion of the issue of clouds is to be found in the book *Chill* (Clairview 2009) by Peter Taylor, a UK environmental consultant of considerable experience and some science training, who has declared himself a sceptic where the science is concerned. Among other items of interest, Taylor points to a host of papers on the amount of radiation reaching the earth in past decades. In particular, he points to a series of papers showing more sunlight reaching the earth's surface in the two decades of warming noted in the previous chapter, and less sunlight reaching the surface in the decade where temperatures seem to have paused, and there may be more on the way. Again this is more to do with clouds than changes in the sun's total output. You will recall in the previous chapter the leaked emails suggesting that climate scientists were plotting how to avoid taking into account various papers that were in the literature. They will have to get busy. The much-vaunted climate models make quite different assumptions about direct radiation from the sun.

Another whole book could be filled with the alleged forecasting successes and failures of the GCMs. The climate models say that there should be a hot spot in the upper troposphere in the tropics which satellites and balloon-borne instruments have not found, and again we are told it's really the instruments that are at fault. The modellers say that they had had some success in modelling ocean temperatures. We will not stop to discuss these points, but again we see the peculiar attitude of climate modellers and their defenders. Problems are brushed aside on all sorts of pretexts such as the measuring instruments are not right, or those making the accusation are biased. In fact, every problem should be sought out, highlighted, examined from all directions, and the models picked apart to see why the result is not what it should be. At no point in the debate have the climate modellers, or even their critics, seem to have grasped that they are dealing with forecasts rather than scientific theories.

The problems are not confined to the gigantic climate models themselves. For the climate models depend on yet another set of forecasts, those of CO_2 concentrations in the air over the next century. The two big greenhouse gases are CO_2, which has by far the highest concentrations of any of the trace gases, and methane (CH_4). Nitrous oxide (think of a car exhaust) some CFCs and other gases are there but collectively make a vanishingly small part of atmosphere. Methane concentrations are also tiny – they are measured in parts per billion, as opposed to parts per million for CO_2 - but as a greenhouse gas it is many times more effective than CO_2, so it has almost one-third of the effect of the more common gas. The two graphs in figure seven for CO_2 and methane concentrations for the last few decades are derived from data collected by NOAA Earth System Research Laboratory in Boulder, Colorado. NOAA stands for the North American Office of Atmospheric Administration.

84 A Guide to Climate Change Lunacy

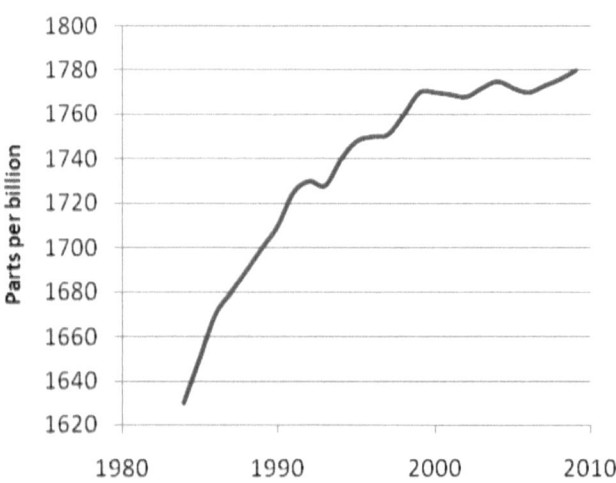

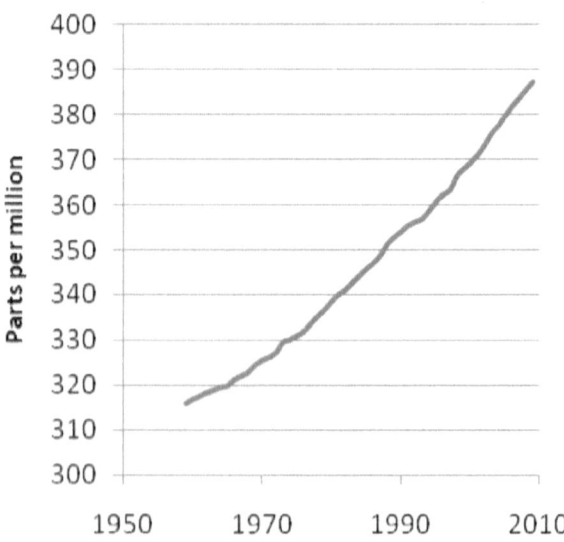

Figure 7: CO_2 *marches ever upward but Methane concentrations have levelled off. No one knows why. Better versions of these graphs can be found on the NOAA site. Source: Mauna Loa and NOAA.*

One piece of good news immediately apparent from the graph on methane is that, despite all the screaming and shouting, alarm and concern over emissions from cows and coal, methane concentrations mostly levelled off of their own accord at about the turn of the century. The IPCC 2007 report gives lots of figures about likely sources of methane including coal production, rice growing, emissions from cows and from the major natural source of wetlands, but has little useful to say about why methane concentrations have stopped increasing. The report tentatively suggests that rice production emissions may have been reduced, or that economic incentives may have curtailed industrial methane production (which does not seem likely). However, as the panel scientists have only recently realised that tropical forests produce methane as well as CO_2, they admit that they do not know enough about the natural methane cycle to be specific.

Then there is carbon dioxide which has been the centre of all the fuss. As you can see results from the site at Mauna Loa in Hawaii show that CO_2 concentrations were 316 parts per million (0.0316 per cent) of air in 1959, and by 2009 the figure was 387ppm. That is an increase of 71ppm in 40 years, or an average of 1.77ppm a year. Those annual increases have accelerated slightly in that time, a point we will get to in a moment, with the main and obvious suspect for that increase being industrial production which expanded greatly in the period, notably in China.

To forecast dire temperature increases the IPCC first has to forecast dire increases in both Methane and CO_2. Most of the modelling work in this regard seems to have been done for the IPCC's *Emissions Report* (or *Special Report on Emissions Scenarios*) released in 2000, which contains more statistics than you will ever want to see on industrial emissions. These are not easy to interpret as the panel's economists forecast emissions in gigatonnes of carbon dioxide equivalent, and the figures that are most useful to us are the equivalent in parts per million (or parts per billion in the case of Methane) in the atmosphere. In addition, the economists give many different scenarios resulting in widely divergent emission outputs. The different scenarios correspond to different forecasts for economic growth, population increases and improvements in technology.

Fortunately tables appended to the IPCC's 2001 report give the output in ppm so it is possible to compare the projections in the 2000 report with the results in 2009 and see that there is a major problem. The lowest projection for Methane for 2010 is 1,839 parts per billion (the forecast is given in 10 year increments), and the highest is for nearly 2000. But the actual result for 2009 as you can see from the graph above was about 1,780. In other words, just 10 years out the projections for methane are wildly wrong. The panel's economists used very sophisticated analytical techniques and economic modelling but, in essence, they expected that the increase in methane concentrations evident in the decades before 2000 would continue and even increase, perhaps because the increase was supposed to be linked to economic activity. Instead the growth in methane concentrations stopped; no-one knows why.

The forecasts for CO_2 concentrations are better, but still short of the horror story the IPCC wants to sell. As we have seen the CO_2 concentrations in 2009 were about 387 parts per million (0.0387 per cent of air), which is a little ahead of the lowest of the IPCC projections I could find, 381ppm. But the mid range projections are mostly above 390ppm and the highest is nearly 400ppm. For just 10 years out that is not very good and again seems to be the result of the panel economists having an exaggerated idea of how industrial production, which has undoubtedly been increasing in that time, affects carbon concentrations. A simple analysis of the Mauna Loa figures giving the annual increase in concentrations with a trend line is given below. Those who do not want to trust figures compiled by 'a mere journalist' (sometimes this phrase is said with a sneer, sometimes it is spat), can always check the material for themselves. It is all available online.

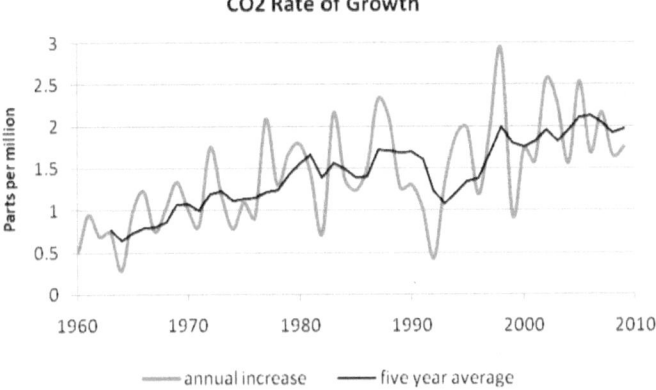

Figure 8: Annual increase in CO_2 concentrations as measured at the Mauna Loa station. The data is readily available for those who go looking for it. The average annual change has shown only a very modest increase for more than a decade.

As you an see from the graph the annual rate of increase in CO_2 has been accelerating in fits and starts since the 1960s. The IPCC economists, working with the best analytical techniques available, seem to have assumed that acceleration would not only continue but also increase sharply. Instead the increase remained steady at around 2ppm per year. Perhaps that is not a fair test as it is only a drop of real world data as opposed to several bucketfuls of the best computer projections available, but it is at least an indication that CO_2 levels will be at the bottom rather than the top of all these projections. The existing CO_2 figures, incidentally, should already reflect the rise and rise of China, particularly as the Chinese have been building an immense number of coal-fired power stations.

There was some opposition to the emissions scenarios when they were released in 2000 from two Australians, Ian Castles and David Henderson. Both men are eminently qualified. Ian Castles is a former Australian Chief Statistician and David Henderson was formerly head of the economic and statistics department at the OECD. As reported by Aynsley Kellow in his *Science and Public Policy: The Virtuous Corruption of Virtual Environmental Science*, the two men objected to the panel's use of currency exchange values rather than Purchasing Power Parity (PPP) as a means of analysis. In the first approach the GDPs of different countries are converted to US dollars at a current exchange rate. In the second different economies are compared with reference to their own internal purchasing power, and without reference to an exchange rate.

One way of looking at this is that PPP may, say, compare the price of a basket of goods in one economy with the same basket of goods in another, expressed as a ratio rather than converted to a common currency. But that is not quite right is that workers in those different economies are not going to be buying the same goods. Workers in one economy will spend more of their salaries on food than those in a more advanced economy. Although a lot more difficult than simply adjusting the GDP figure to US dollar values, PPP analysis is thought to be a considerably more accurate measure.

In a 2005 paper in *Energy & Environment* entitled *SRES,* 'IPCC and the Treatment of Economic Issues: What has Emerged?' Henderson gives the example of the GDP of Brazil which grew by an estimated 13 per cent over the period 2000-2004. But when measured in market exchange terms the Brazilian economy appears to have shrunk by 43 per cent, simply because of the change in the exchange rate at the time. The market exchange rate (MER) analysis gave a completely wrong answer. In the paper he says that other UN agencies have agreed to use PPP analysis, but the IPCC economists persist in considering MER and PPP as two different ways to approach the same problem. He says with some force that they are not. The PPP approach is valid; the MER approach cannot be used.

A major result of using MER is that the economies of developing countries in particular are substantially understated at the beginning of the scenarios, as their exchange rates are lower. But various IPCC scenarios assume that the economies of those developing countries will be up near OECD levels (and exchange rates) by 2100. In other words the economies of those countries are set up to be small at the beginning of the scenarios and very rich at the end, with corresponding increases in emissions. Kellow says that the Castles and Henderson critique drew favourable responses but the IPCC responded by trying to discredit the pair. In his 2005 paper, Henderson reproduces an IPCC press release describing the dispute as "disinformation"

In chapter three of the 2007 report the IPCC agrees that PPP should be used where practical (some of the chapter's material is expressed in PPP), but says that the relevant data is not available for a number of the economies in the forecast. The panel's economists also repeated the point that they do not think the end result will be any different.

Despite this controversy, the fact that CO_2 and methane concentrations have fallen short of expectations; and that the emissions report itself says that all the scenarios are equally likely, other groups persist in repeating the very worst of the forecasts issued by the IPCC. Australia's Garnaut Report, released in 2008, discussed in a little more detail in the chapter on the economics of climate change includes methane and other industrial gases in a mix which it says will increase by an additional 100ppm by 2030. This would require growth rates in CO_2 concentrations two and a half times the rates now being measured – rates that have been steady for a decade. When I first saw the Garnaut figure I thought it must be a misprint. The Monaco Declaration discussed in the chapter entitled 'Acid and Adaption' says that CO_2 concentrations of 560ppm are possible by mid century, which is also ludicrous. The scientists and economists in both groups are probably using the 2000 projections and trusting to breathless pronouncements that industrial emissions are higher than expected. Industrial emissions may be higher for all I know but the extra emissions have yet to show up in the actual CO_2 concentrations which, in theory, are meant to determine temperatures.

Statements made at the Copenhagen Conference in December 2009, which activists hoped would lead to effective, enforceable limits on emissions, also indicate that only the top most or worst of those 10 year old projections are being used. The old projections also seem to have been used for the climate models in the 2007 report. No-one has thought to recast these projections with regard to 10 years of actual results.

One answer from the panel to all this wearying criticism is to point to what amounts to a carbon dioxide hockey stick. The IPCC 2007 report states that the acceptable pre-industrial value for CO_2 is 275 to 285ppm. It took 200 years to increase that level to 336, but only 30 years to increase concentrations by another 50ppm, and never mind that the last few decades have shown only comparatively minor changes in the rate of increase. Unfortunately for this argument the panel's nominated value for pre-industrial levels – that is the level of CO_2 in the atmosphere before people started building a lot of factories – has also come under attack.

Although ice core measurements of the last few centuries show CO_2 concentrations as a dead flat line at about 275-285ppm, there are indications that those measurements skip a lot of changes. Scientists have found that they can reliably measure CO_2 concentrations in medieval times by counting the number of stomata (gas exchange pores – that is the leaf's mechanism for breathing air) on leaves of English oak trees preserved in peat bogs. Dutch scientists managed to find and examine enough leaves to conclude that pre-industrial CO_2 levels varied by perhaps 20 parts per million around an average level of 311ppm in medieval times. That is from 290 through to 330ppm. 'A role for atmospheric CO_2 in pre-industrial climate forcing', B van Hoof et al, (*Proceedings of the National Academy of Sciences of the USA*, 14 October 2008; available online).

Another paper by Friederike Wagner, of the Department of Botanical Palaeoecology at Utrecht University in the Netherlands, uses similar techniques from much older leaves to get natural CO_2 concentrations of up to 325ppm from well back in the *Holocene*. 'Rapid atmospheric CO_2 changes associated with the 8,200-years-B.P. cooling event', (Proceedings of the National Academy of Sciences, 17 September 2002.) As CO_2 concentrations can vary naturally up to 330ppm plus in warm times, at least in this method of measuring, this much higher level of natural CO_2 is also possible in modern times. However, that higher level messes up the pure shape of CO_2 hockey stick and puts a question mark over the painstakingly collected CO_2 readings from ice cores. They may be consistently too low, and certainly miss out on a lot of detail. The Wagner paper also points out, in passing, that ice core results from Greenland have shown CO_2 readings well above 400 ppm but the results from Antarctica are always preferred.

Another crucial point mentioned in both papers is that there is a correlation between CO_2 levels and ocean temperatures, where those temperatures have been established by different means. As we have already discussed warmer water expels CO_2 and cooler water can hold more of the gas (think of warm soda water), and it is widely agreed that the oceans hold sixty times more CO_2 than the atmosphere. There is some confusion here. As we have seen, the ice core measurements also show that CO_2 levels follow sea temperatures although with a lag time that can be more than 1,000 years, but the leaf stomata work indicates that CO_2 levels follow temperatures almost in lock step. Which is right? One possibility is that as the ice core results consistently underestimate CO_2 concentrations, perhaps they do not show changes in CO_2 until well after they started happening? Or perhaps the leaf-reading results are wrong, or perhaps both are wrong. Remember that we are seeing the current CO_2 concentrations in high resolution with at least consistent measurement

Here we can quickly get caught up in an argument over the way CO_2 levels have been measured in ice cores and the way concentrations in the atmosphere were measured before the modern measuring networks were set up in the 1950s. Zbigniew Jaworowski, a Polish scientist who is now a senior advisor at the central laboratories for Radiological Protection in Warsaw and former chair of the United Nations Scientific Committee on the Effects of Atomic Radiation, but with considerable experience in ice core measurements, has been pointing to the fossil leaf indices for some time. He also says there are flaws in the way that ice core CO_2 concentrations are calculated, and that before 1985 CO_2 concentrations measured in ice cores ranged from 160 to 700ppms. After 1985 those high readings disappeared, although the ice cores results cover intergalacials that are known to be much warmer than modern times.

Although Jaworowski has some backing from the fossilised leaf work, global warming scientists have done their best to ignore him, as well as ignore the larger number of scientists who have complained repeatedly about analysis of CO_2 measurements taken before the modern instrument network. The key piece of evidence here is a controversial paper by Ernst-Georg Beck, a biologist at the University of Reiburg ('180 years of Atmospheric CO_2 Gas Analysis by Chemical Methods', *Energy and Environment*, 2007). Atmospheric CO_2 concentrations were measured lots of times in the later half of the 19th century and those readings can be analysed. Beck says that, in effect, earlier analyses of these measurements discarded most of the high readings to arrive at the magic figure of 280 ppm. He puts them back in the mix to derive a curve for CO_2 which follows what is known of temperatures at the time, but with a lag of about five years. He also estimates that CO_2 levels reached 410ppm in 1940 before falling to the starting point of the Mauna Loa readings!

The paper has since been bitterly attacked by global warming scientists and activists, with one response being that they do not see how so much carbon could get into the air from vegetation and then be removed (the ocean is both the source and sink). Although the fossil leaf work indicates that Beck may have at least a partial point – a lot of variation is not picked up by the ice core work - the concentrations in those papers are still much lower than anything found by Beck. However, the levels of pre-industrial CO_2 concentrations remains one of those issues that global warmers find so irritating but cannot get rid of. Every now and then a scientist such as Joel M. Kauffman, an emeritus professor at the Department of Chemistry and Biochemistry at the University of the Sciences in Philadelphia, puts his head above the parapet to say that there may be something in Beck's paper. ('Climate Change Re-examined', an essay, *Journal of Scientific Exploration*, 2007). Another point to note is that methane concentrations levelled off at about the same time as temperatures, which was also about the same time that increases in CO_2 concentrations settled down.

That is all far from definitive. In fact, as far as the mainstream are concerned Beck and co are the cranky fringe, but we should keep the above discussion in mind in our wander through the next and strangest of all parts of this long, bizarre trip through the "settled" science of global warming – that of the isotopic signature of carbon dioxide in the atmosphere.

The IPCC lays the groundwork in the 2007 report's chapter two on 'Changes in Atmospheric Constituents in Radiative Forcing, section 2.3.1 Atmospheric Carbon Dioxide'. In that section panel scientists write that the bulk of the carbon in carbon dioxide is made up of the ^{12}C isotope, that is the "normal" carbon atom with six protons and six neutrons in the atom nucleus. But one per cent is ^{13}C which has an extra neutron. The atoms are chemically the same, just one is slightly heavier than the other. This ratio can change depending on the source and, crucially, the carbon in fossil fuels has a slightly different ratio to that of the carbon in natural CO_2. It has a little less of the ^{13}C isotope, which is usually just 1 per cent of carbon to begin with. This ratio can be measured and there is an evident trend consistent with increasing CO_2 production from fossil fuels. The IPCC constructed a graph and added in a line showing increasing industrial emissions. The graph from the IPCC 2007 report is reproduced below.

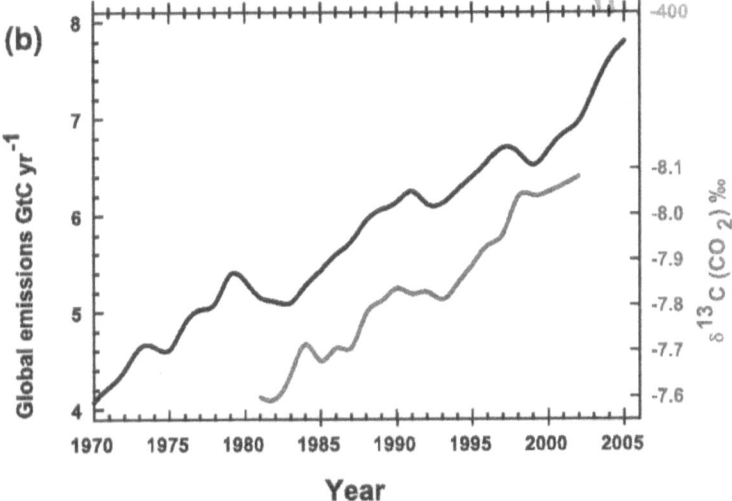

Figure 9: Trends in Carbon Isotopes in the Atmosphere, Source: IPCC.

The global emissions are given on the left of the graphic in gigatonnes of carbon per year, shown by the top wavy line. The bottom wavy line (in red in the IPCC report) is the trend in the isotope ratio, and is measured on the right hand scale. The strange symbol is a delta sign and just means change in the ratio of ^{13}C to total CO_2. This ratio (denoted by the per cent figure with the extra zero on the side) is compared with a standard ratio which scientists have previously agreed upon. The actual number is not so important, it is the trend and how it compares to other numbers that matters, and scientists should be able to read that signature to determine what proportion of CO_2 now in the air has come from fossil fuel sources. Unfortunately, the answer is not one the global warmers will like at all.

The IPCC says nothing whatever about the ratio beyond stating, correctly, that it is an indication of more fossil fuel emissions in the atmosphere. Tom V. Segalstad of the Mineralogical-Geological Museum, University of Oslo in Norway, however, points out that the ratio implies that just 4 per cent of the CO_2 in the atmosphere comes from fossil fuels, and not the 21 per cent claimed by the IPCC. In an essay in the *The Global Warming Debate – The Report of the European Science & Environmental Forum*(Bourne Press, 1996) Segalstad, a professor of resource and environmental geology and a former head of the Natural History Museum and Botanical Garden at the university says that an entirely natural atmosphere would have a ratio of about -7. If 21 per cent of the CO_2 is from fossil fuel then the ratio should be about -11. Assuming the ratio varies in a straight line then the 8.1 ratio above implies that fossil fuels make up just 5 per cent of the atmosphere. (The four per cent figure is from a decade or so ago.)

Segalstad has repeated this point many times and in a 1998 report by the same group cited above, *Global Warming – the Continuing Debate*(both chapters are available on Segalstad's web site, www.co2web.info), he has many harsh things to say about the IPCC's attempts to model the amount of time CO_2 lingers in the atmosphere before being absorbed by one means or another. He cites 35 refereed papers published over more than three decades up to the early 1990s that give an average CO_2 residence time (the time it hangs around in the atmosphere) of five to 10 years, mostly from measurement. The maximum estimate is 15 years.

The IPCC approach to working out CO_2 residence time is completely different. Chapter seven of the 2007 IPCC report, in the section entitled 'Are the Increases in Atmospheric Carbon Dioxide and Other Greenhouse Gases During the Industrial Era Caused by Human Activities?' refers to a "Bern Carbon Cycle Model" and states that about 50 per cent of an increase in atmospheric CO_2 will be removed within 30 years, a further 30 per cent will be removed within a few centuries and the remaining 20 per cent may remain in the atmosphere for many thousands of years. The IPCC also notes that natural sinks take up nearly half of the human produced CO_2, and discusses various sources and sinks at considerable length in chapters two and seven. A glance at that material shows that the human generated CO_2 is small compared to the ebb and flow of natural CO_2 (a commonly cited estimate is about 2 per cent) but the fossil fuel CO_2 addition is supposed to be crucial.

Segalstad says that if the IPCC goes back to the original, well established figure for CO_2 residence time then emissions line up with the nuclear signature data above. In effect he says that the IPCC has assumed that the bulk of the growth of CO_2 from pre-industrial levels is due to fossil fuel use, and had to re-estimate the residence times using computer models to make that assumption work. The resulting residence times proved too long, as they put too much CO_2 in the atmosphere, so the panel declared that half of industrial emissions were absorbed while the rest lingered.

Segalstad has made the same points many times in recent years. In an interview with the Canadian paper *Financial Post* (7 July 2007) he says that the IPCC scientists have not tried to overturn or contradict the earlier studies on CO_2 residence times, they have just dismissed them and relied on computer models. He also says they have no measurements or physical evidence to back up their claims that CO_2 lasts decades or centuries in the atmosphere.

These are far reaching claims which imply that only (about) seven years' worth of fossil fuel emissions are relevant at any one time and that the bulk of the CO_2 increase in the atmosphere is natural. In this reading of the science only about 20ppm of current concentrations are due to human activity. Even if we adopt the fossil leaf results as a natural baseline that still leaves a lot of unexplained CO_2 in the air, but you will recall that we are seeing modern CO_2 changes in high resolution, with reliable, consistent measurement. This debate has also proved so strange that I have ceased to be surprised by any of its turns. In this version of the science CO_2 levels rely mostly on sea surface temperatures, so when the oceans cool as they are expected to do then CO_2 levels should start to level out and then fall, perhaps after a lag of seven to 10 years. Cannot possibly happen? Rather than argue about the theory why don't we wait and see?

Despite doing his best to be vocal, Segalstad has yet to face any serious counter-argument. Greenhouse scientists have not sneeringly dismissed them, or disagreed with him, or given any reasons why they have ignored the earlier work on carbon dioxide residence times. They have said nothing at all; not even pointing out, with some truth, that using these ratios for absolute measurement rather than just as a way of comparing trends, can be difficult. Even Wikipedia, which can always be relied on for a lengthy justification of greenhouse science, says only that he disagrees with mainstream science. Segalstad told me, by email, that the issue hasn't been discussed in the literature.

In legal matters, silence does not mean much apart from the suspect not making any admissions, but in the greenhouse debate silence is suspicious. Considering just how bitterly global warming scientists attack unbelievers, my interpretation of the silence is that they do not want to draw attention to arguments for which they have no answer. Hardly anyone seems to have supported Segalstad but, as we have seen, numbers of scientists for and against a particular proposition are essentially meaningless in adjudicating scientific realities. There is one straightforward way to work out who is right and wrong in this, however, and that is see what CO_2 emissions will do.

One answer sometimes given by those stuck for any other counter-argument to the atomic signature material is that not referring to it implies a conspiracy among scientists. Not at all. Peer group pressure is very strong among scientists. This is why theories persist for many years despite contrary evidence, as we have seen. Occasionally someone sticks their heads above the parapet, but it does not happen as often as it should. There is also the not incidental question of funding. Asking difficult questions about issues with which the scientists are not directly concerned may threaten what has proved to be a lucrative source of funding. Best to stick to their own box.

Before moving onto the arguments over the last decade's worth of temperatures, and the much-derided solar magnetic model of climate change, we should look at efforts to use the IPCC carbon-based climate models to investigate climates far more ancient than the ones covered by the graph at the start of the chapter on ice ages. These efforts have achieved, at best, mixed results.

In a recent paper in *Nature Geoscience* three American scientists admit that they cannot work out what happened at the Palaeocene-Eocene Thermal Maximum (about 55 million years ago), when a lot of carbon got dumped into the atmosphere and temperatures rose 5-9C in a few thousand years.

The abstract for the paper by Richard E. Zeebe of the University of Hawaii and others says: *"At accepted values for the climate sensitivity to a doubling of the atmospheric CO_2 concentration, this rise in CO_2 can explain only between 1 and 3.5 °C of the warming inferred from proxy records. We conclude that in addition to direct CO_2 forcing, other processes and/or feedbacks that are hitherto unknown must have caused a substantial portion of the warming during the Palaeocene-Eocene Thermal Maximum. Once these processes have been identified, their potential effect on future climate change needs to be taken into account."* ('The Carbon dioxide forcing alone insufficient to explain Palaeocene–Eocene Thermal Maximum warming', *Nature Geoscience*, 13 July 2009).

Before the event, CO_2 levels were around 1,000ppm, or a little under three times present levels, with a climate much warmer than present times. The group estimates that during the main phase of this event CO_2 levels rose to 1,700ppm. The carbon orthodoxy has been kept alive by speculation that some warming may have driven a lot more methane out of swamps, which caused more warming.

However, it remains an inconvenient piece of research for global warmers, and is just one of a series of difficulties encountered by the carbon-based models in investigating any of earth's known prehistoric climates.

An article in *New Scientist* 'When crocodiles roamed the Arctic' (18 June 2008) discusses the period 100 to 40 million years ago when the Antarctic was covered by almost sub tropical forests. At the same time the Arctic, largely isolated from the other oceans at the time, was a gigantic freshwater lake filled with crocodile-like creatures. Scientists have gone to some trouble to use models to simulate these conditions, helped by the fact that it is hard to measure CO_2 levels that far back so they have some freedom in setting levels. The article says they have used levels eight to 16 times higher than industrial times. Eight times industrial levels works out to around 3,100 parts per million, which is far above any previous estimates of CO_2 concentrations of the time or the estimates in the *Nature Geoscience* paper cited above.

After some theoretical pushing and shoving modellers managed a partial fit with those very high CO_2 levels, but were still left with too much heat at the Poles so they added in a lot of hurricanes to transport the heat out. That still leaves them with the puzzle of why the tropics of the time were not unbearably hot and the high land regions such as Siberia very cold, as the models say they should. Perhaps in time scientists will make their models fit that period.

But let us move on. There is yet more lunacy to come.

5
CONFUSION PAST AND PRESENT

As we saw in previous chapters the last decade or so of temperatures has proved disappointing for those hoping for a climate catastrophe to prove the inherent dangers of industrial activity. Towards the end of 2009 global warming groups reluctantly began to admit it, with those admissions revealing distinctly different views of the science. In addition, those explanations include an extraordinary admission by an ardent pro-warming scientist that the decline in temperatures may continue for decades thanks to a major oceanic cycle going into a cooling phase.

Noel Keenlyside of the Leibnitz Institute of Marine Science in the German city of Kiel was one of the first to admit there was a problem. As reported in the journal *Nature* ('Advancing decadal-scale climate prediction in the North Atlantic sector', *Nature letters*, 1 May 2008) Keenlyside and colleagues note that an oceanic climate cycle known as the Atlantic Meridional Oscillation (AMO) was weakening to a mean and that weakening would result in a cooling effect. So they added the effect of the AMO cycle to the IPCC models to forecast that global temperatures will remain stable or perhaps even dip down for the next decade, before heading up. In a subsequent interview with the *Daily Telegraph* in the UK Keenlyside stated that the earth will start to warm again in 2015. In his public comments he has also emphasised that his work in no way contradicts that of the IPCC; he is merely adding climate cycles on top of the panel's predictions.

The item in *Nature* says that sea surface conditions in the Atlantic are known to influence hurricane activity, as well as surface temperature and rainfall variations over North America, Europe and northern Africa. Those variations may be predictable if the current state of the ocean is known, but the lack of subsurface ocean observations (temperatures and conditions at different depths) remains a problem. Researchers can partially overcome the problems with sea surface temperature (SST) observations, and are improving their skill, indicating that routine decadal climate predictions may be possible. All this sounds quite hopeful and shows that a lot of good may come out of climate research, but it is not what global warmers want to hear. Where were the warnings about tipping points just around the corner? Talk of cooling detracts from the message! There were reports Keenlyside was challenged to bets.

In 2009 an Australian Senator for the Family First Party, Steve Fielding, triggered another admission when he returned from a conference in the US to declare that some scientists were saying earth's climate is driven by "solar flares" (actually he meant the solar activity or solar magnetic activity theories which will be discussed briefly below). He convened a debate between sceptical scientists and those who supported the government's climate legislation. We will not rehash that debate but a part of the response by Environment Minister Penny Wong written by pro-warming scientists is worth noting.

They say: *"The observational evidence clearly indicates that the climate system has continued to warm since 1998. During this period ocean heat content has risen, ice and snow have continued to melt, and there has been no material trend in global air temperatures. When changes in surface air temperature are considered, it is important to note that at time scales of around a decade, natural variability can mask the atmospheric warming trend caused by the increasing concentration of greenhouse gases. For example, global average surface temperatures clearly increased between 1975 and 2008 but some shorter periods, such as 1981-1989, showed no warming. Such behaviour is consistent with the outputs of climate models such as those assessed by the IPCC."* (Senator the Hon. Penny Wong, Minister for Climate Change and Water, response to Senator Fielding's questions about the climate change science).

Now instead of a change in ocean cycles masking the warming, the heat is going into the oceans. Also note that in the decade chosen as an illustration of a temporary lack of warming, 1981-1989, temperatures were increasing. They have been declining since 1998 (the response says "no material trend").

The response includes some graphs which supposedly show that the oceans are warming but are unsourced. The sceptics later grumbled they had to guess which paper the graph came from and pointed to three other papers which don't show warming in the oceans. The argument need not detain us, because the pro-warming scientists are quite right to point out that natural variations over a decade can mask other forces at work but, if so, what causes those natural variations? When are they likely to end? How do we know that the increase between the mid-70s and the beginning of the century was not also a natural variation? The IPCC, incidentally, declared in its 2007 report that all the global temperature changes up to 1940 are natural. The report does not give any reason for this declaration, although it is probably because CO_2 increases in the atmosphere were not considered significant until then. It just blandly states that point and moves on.

The pro-warming scientists are also right to point out that the oceans are the key to climate but rather than bother with the details, let us look at another explanation of the pause in temperatures from the pro-warming side.

About the same time as the Minister for Climate Change's scientific team was working itself into a lather over ocean temperatures, Judith L. Lean, of the Naval Research Laboratory in Washington and David H. Rind of NASA's Goddard Institute for Space Studies, a bastion of the IPCC, published a different explanation. In a paper forecasting temperature increases for the next two decades, 'How will Earth's surface temperature change in future decades' (*Geophysical Research Letters*, 15 August 2009), they suggest that the earth will be warming at twice the rate forecast by the IPCC, but it also concede that natural variations will mess up the forecast. In discussing a particular part of the forecast period when there is less warming they say, "This lack of overall warming (in the forecasts) is analogous to the period from 2002 to 2008 when decreasing solar irradiance also countered much of the anthropogenic (artificial) warming". In other words, according to this lot the sun is the problem. They don't say anything about oceans. Also note the casual way the point is addressed, as if it's almost not worth mentioning. Like the Australian government scientists, the US researchers may have a point, but the favoured explanation remains that of ocean cycles disguising the warming.

In September 2009, Mojib Latif of the Leibniz Institute of Marine Sciences at Kiel University, Germany, a climate modeller and definitely not a climate sceptic, conceded that part of the warming between 1975 and the end of the century may have been due to the oceanic-atmospheric cycles known as the North Atlantic Oscillation (NAO) and the Atlantic Meridional Oscillation (AMO). In a speech at the World Climate Conference in Geneva in September called by the World Meteorological Organisation, Latif also concedes that as these climate cycles are in a "cooling phase" temperatures will decline for the next few years ('World will cool for the next decade', *New Scientist*, 9 September 2009). In the same article, Vicky Pope, head of climate change advice at the UK Met Office agreed that natural variability was important in the short term, and that the office knew more about what would happen in 2050 than next year. In other words the forecasts are wrong in the short term but, she contends, still right in the long term. We shall see.

Other prominent scientists have joined the chorus. Kyle L. Swanson of the University of Wisconsin-Milwaukee also discussed climate variability in the paper 'Long-term natural variability and 20th century climate change' (*Proceedings of the National Academy of Science*, online 14 September 2009.) In a special supplement to the August 2009 Bulletin of the American Meteorological Society, *State of the Climate in 2008*, Dr Jeff Knight and colleagues from the UK Met Office Hadley Centre say that the for the past 10 years the HadCRUT3 temperature data show that the world warmed by 0.07 degrees but with and error of +/- 0.07 – in other words no warming at all. But the *State of the Climate* supplement expects warming to come back.

No one stopped activists at the conference in Copenhagen in December 2009 from screaming that tipping points are fast approaching, and the hard-liner scientists, who are also the most vocal and active in pushing the greenhouse line, are not about to agree to anything. But there does seem to be a tacit agreement that we will have to wait a few more years for our climate catastrophe, and perhaps a few more years after that.

Other climate cycles are following the AMO lead. Or perhaps the AMO is following those cycles, as comparatively little is known about the cycles in general or the links between those in different ocean basins. In April, NASA announced that the Pacific Decadal Oscillation has shifted from its warm mode to its cool mode - 'Larger Pacific Climate Event Helps Current La Niña Linger' (NASA – JPL, 21 April 2008). PDO shifts from cool to warm phase and back again every 20 to 30 years. During most of the 1980s and up to the mid-1990s, the Pacific was locked in the oscillation's warm phase, with cool water in a horseshoe shape around the north, west and southern Pacific and warm water in the middle. In the cool phase, the cool water is on the outside and the warm water in the middle.

The PDO interacts with the already mentioned El Niño/Southern Oscillation (ENSO). When the PDO is in a warm phase the El Niño effect is amplified, as happened in 1987-88. When it is in its cool phase the La Niña part of the ENSO cycle may be amplified, as it was in 2008-09, and the El Niño part is subdued. As this book is being written in early 2010, as El Niño event rules but it is proving to be subdued. In this view of climate the PDO and AMO direct major trends in climate, while the ENSO cycles cause minor changes.

As the PDO itself is a comparatively recent discovery, the term was only coined by American fisheries scientist Steven Hare in 1996 when he noticed a link between Alaska salmon production cycles and Pacific climate, it is difficult to say more. But one scientist who has made forecasting climate change from the PDO his particular bailiwick is Don Easterbrook, a professor emeritus of Geology at Western Washington University. In 2001, a few months after the IPCC delivered its report forecasting run away warming, Easterbrook pointed to a distinct connection between the cool and warm phases of the PDO and cooling and warming of the climate. In a paper presented to the Geological Society of America's annual meeting he forecast global cooling for the next 25 years. ('The next 25 years: global warming or global cooling? – geologic and oceanographic evidence for cyclical climate oscillations'.) The PDO was in its warm phase between the mid-70s and the late 90s, so global temperatures increased. It was in a cool phase between the mid-1940s and the mid-70s so the earth's climate generally cooled (the same argument is used for the AMO above – so are the two cycles linked or what?). In other papers he says that the cycle can be traced back and in the past century each phase has been 25 to 30 years long. Now that the PDO has switched to its cool phase he says that we can expect 25 to 30 years of cooling. Very well, but why should we listen to this one scientist. The reason we should pay attention to Easterbrook over the mighty IPCC is that his forecast was better than the panel's. Therefore Easterbrook wins and the IPCC loses, at least for the moment. Let's see what happens.

There is a great deal more to oceanic cycles and currents, including a whole system of deep water circulation that is only vaguely understood, but it seems that the earth is going to cool for a while. Global warming scientists insist that this cooling is only temporary and that the insidious warming from industrial gases will eventually win out, but that depends on their beloved climate models being of any use. As we have seen, there is a very good chance they are wrong in many ways. In particular, there is a good chance that the models are missing a very important piece of the climate puzzle – solar activity – with solar activity being the cause of the changes in ocean cycles, at least as far as scientists know at this point.

The sun obviously has a direct influence on climate, through changes in total solar output, but it may well have a more subtle influence through its magnetic field and the streams of charged particles it sends out. These are always changing. Scientists are only just starting to tease out these complications, much to the disgust of the global warming scientists, but there is already strong historical evidence of a link.

In the first chapter we discussed sunspots which have a marked 11 year cycle. At the top of the cycle when there are lots of spots on the sun, its magnetic field changes, lots more particles stream out from the star and it emits solar storms that upset satellites. The sun is said to be active although its total output does not change all that much. The change is in its magnetic field and in the stream of particles, mostly protons and electrons, it sends out in a thin, charged plasma called the solar wind (scientists even talk of solar weather although its still a vacuum). At the bottom of the cycle there are few or no sun spots, much less magnetic activity and the "wind" is reduced. Scientists have long pointed out that the Maunder minimum – a period of decades when the sun showed no sunspots – more or less coincided with the worst of the Little Ice Age, and the connection does not end there.

Systematic observations of sunspots only started in the 17th century but scientists have been able to reconstruct a solar activity index back 10,000 years. A fact sheet produced by the US Department of the Interior and the US Geological Survey, 'The Sun and Climate' (USGS Fact Sheet FS-095-00, August 2000), says that solar activity can be tracked through the production of isotope of carbon with an atomic weight of 14 (we discussed isotopes at the end of the chapter on ice age escapades). This isotope is produced in the upper atmosphere as part of a complex process involving cosmic radiation – an endless stream of protons hitting the earth's atmosphere. Space is a busy place it seems, despite being a vacuum.

The intensity of cosmic radiation is influenced by solar magnetic activity. When the sun is active, as indicated by the number of sun spots, the more active magnetic field shields the earth from cosmic radiation. More solar activity, fewer neutrons and so less 14C. Less solar activity, more neutrons and so more 14C.

That additional 14C can be found in minute quantities in the wood of trees. This is convenient as the trees also have tree rings where each ring corresponds to a year, so it is possible to calculate dates which can then be matched to the 14C readings. Trees don't last thousands of years but by finding bits of wood that can be radiocarbon dated, or even fossilised trees from which they can somehow extract measurements, scientists have been able to construct an index of solar activity back 10,000 years. The index is an excellent match for what is known of climate history, with a mid-holocene maximum about where it should be, as is the Roman Warming, Medieval Maximum, Little Ice Age and modern warm period. The match is so good that even Wikipedia, that bible of global warming orthodoxy, grudgingly admits that there is some correlation, with the admission buried in an article on solar variation. However, the correlation is supposed to have broken down in the last 30 years or so thanks to evil industrial activity, a point we shall come to in a moment.

This correlation was confirmed by others and taken further in a speculative but entertaining paper by Charles A. Perry, of the US Geological Survey, and Swiss scientist Kenneth J. Hsu. The paper, 'Geophysical, archaeological and historical evidence support a solar-output model for climate change' (*Proceedings of the National Academy of Sciences*, 7 November 2000; available online). The paper presents evidence of larger and longer cycles in the sun than just the 11 year basic sunspot cycle, and that the cycles fall into a noticeable pattern. The scientists then extrapolate that work to suppose that there are 13 individual cycles, ranging from 11 years through to 90,000 years, or the length of an ice age.

Using the end of the last ice age as an anchor point, they combined all those cycles to give a pattern which follows the highlights of climate history including the Holocene Maximum and Medieval Warm period. The warm periods are when one or more than one cycles are up, cold periods are when they are down. Far more importantly, the analysis also forecasts a sharp fall in solar activity off a peak in about 2000, and that seems to have happened, albeit a decade later which is not bad considering the scale of the work. In earlier chapters we looked at a major climate cycle that is known to have a period of 1,500 years, give or take 500 years, and also runs for hundreds of thousands of years. The previously mentioned Singer-Avery book *Unstoppable Global Warming* suggests that the 1,500 year cycle is the result of two interacting solar cycles. Perry and Hsu's work also indicates that the present interglacial, which has proved so comfortable for humans, will last another 10,000 years with occasional bouts of colder climate.

Despite the research being around for a while and the various climate cycles being well known, it was all ignored by global warmers until a link was identified – a possible mechanism by which the sun may cause these climactic sea-saws – and the theory got some publicity. In 1995 Henrik Svensmark at the Danish Meteorological Institute built on earlier work by Knud Lasson and Eigil Friis-Christensen at the same institute by showing a correlation between the extent of cloud cover and relative intensity of cosmic rays. As we have seen a more active magnetic field helps shield the earth from cosmic rays. Fewer cosmic rays means fewer clouds are formed, and a reduction in cloud cover means a warmer earth. More cosmic rays means more cloud, and those clouds cool the earth. As you will recall in the previous chapter there was other evidence of more sunlight during the warming burst that ended around 2000, which implies less cloud, and some evidence of reduced clouds during the same period.

In 2005 Svensmark was able to show experimentally that the shower of particles generated when cosmic rays hit the atmosphere could seed clouds, and his ideas were given considerable publicity on the 2007 programme the *Great Global Warming Swindle* made by Martin Durkin for Channel Four in the UK. For a full discussion of the this programme and the reactions I can only refer you to others, such as Christopher Booker's *The Real Global Warming Disaster* who have already covered it extensively, but one notable response was a paper designed to counter the solar magnetic argument by Mike Lockwood, a physicist at the Rutherford Appleton laboratory and Claus Frölich, of the word radiation centre in Davos, Switzerland.

The paper, 'Recently opposite directed trends in climate forcings and the global mean surface temperature ' (*Proceedings of the Royal Society A*, 13 July 2007) did not try to contest the evidence that the sun and climate were strongly linked in pre-industrial times and even in the first half of the twentieth century. In fact, the scientists use the first two pages of their paper to acknowledge the mass of evidence for the sun's influence on climate, including the key work of Gerard Bond of the Lamont-Doherty Earth Observatory of Columbia University and nine others which convincingly linked solar activity to changes in drift ice in the North Atlantic over thousands of years ('Persistent Solar Influence on North Atlantic Climate During the Holocene', *Science*, 7 December 2001). But after waving around a lot of graphs and numbers the pair conclude that the link breaks down in 1985, with the implication being that the warming after that must be due to industrial gases.

Global warming supporters wasted no time in trumpeting the paper as convincing demolition of the solar-link theory, ignoring all the counter-arguments to Lockwood's paper. Unfortunately for their case, the bare admission that a link existed up to 1985 does untold damage to the global warming theory. If such a link exists – and the paper does not say very much about the link itself - why have we, and the IPCC, been messing around with hockey sticks and arguments over past climates? Why are scientists still poking and proding the Milenkovitch cycles and why are geologists producing papers which "prove" that carbon dioxide has really been driving climate change throughout geological time.

In doing away with all those very clever theories the solar theory also eliminates a lot of the problems. At the end of the ice age increases in magnetic activity in the sun shielded the earth from cosmic radiation, and so greatly reduced cloud cover to let in more sunshine, among other effects. The earth warmed and the warming ocean gave off CO_2, just as warm soft drink loses it fizz. CO_2 concentrations follow temperatures, perhaps more closely than previously thought. The solar model is also a much better fit for the abrupt temperature changes at the end of the intergalacials than the slow moving Milankovitch cycles, and for changes in climate during the holocene.

However, the solar theory also cuts at the heart of one of the main justifications of global warming theory. When scientists account for all the known major drivers of climate, such as solar radiation (as opposed to magnetic activity), aerosols, industrial emissions, ozone and albedo effects, they point out that only carbon dioxide has increased at the same time as recent temperatures increased, therefore the gas must be to blame. That reasoning entirely overlooks solar magnetism which may be the main driver of climate. Never mind whether anyone thinks that its effects have been swamped by other factors in recent decades, the solar magnetic influence still has to be assessed, its effects calculated, the results put into the models and the whole exercise redone. Any forecasts made with the models are, of course, now invalid.

There are other implications. You will recall that the much vaunted hockey stick graph retired hurt in 2006, leaving the global warmers with no direct proof that industrial emissions caused any part of the global warming that has occurred in the past century and a half. They replaced it with an argument concerning the models which is best explained by a senior and distinguished scientist on the global warming side. In late 2009 a serious blog on public affairs called The Fort Collins Forum based in Colorado in the US, hosted a debate on climate between scientists William Gray and Kevin Trenberth who both live in the area. The previously mentioned Gray is Professor Emeritus of Atmospheric Science at Colorado State University and Trenberth is head of the Climate Analysis Section of the National Centre for Atmospheric Research in Boulder.

We will not go over the details of the debate but point to one part of Trenberth's response, which is on the usefulness of models. "Today's best climate models are now able to reproduce the observed major changes of the past century. When the models are run without human changes in the atmosphere, the natural forcings and intrinsic natural variability fail to capture the increase in global surface temperatures over the past 35 years or so. But when anthropogenic (human induced) changes are included, the models simulate the observed global temperature record with impressive accuracy. Observed changes in storms and precipitation are also replicated only by models with human changes in atmospheric composition."

In other words scientists have taken climate models that still have no forecasting track record to speak of, or been verified in any accepted sense, and used them to demonstrate that industrial gases are affecting climate. As we have seen, those models already contain major assumptions about the effect of industrial gases, so they are being asked to prove what has already been programmed into them. You will also note that Trenberth is only claiming success in this back testing exercise for the past 35 years or so.

This very doubtful "proof" using computer models is also cited, in all seriousness, in *The Science and Politics of Global Climate Change* by Andrew E. Dessler and Edward A. Parson (Cambridge University Press, 2006), using graphs taken from the IPCC 2001 report. The book by Dessler, an associate professor in the Department of Atmospheric Sciences at Texas A&M University and Parson, an associate professor of Natural Resources and the Environment, was in its ninth printing at the time of writing. Nor does it stop there. I have sighted this circular logic in at least two articles in top-line journals.

The scientists quoting this form of proof are obviously highly intelligent and with qualifications far above that of any humble layman journalist, but it is not a matter of staring open mouthed at such clever arguments thinking that because distinguished scientists are presenting them they much be right. It is a matter of staring open mouthed at what is an obviously circular argument and asking why wasn't this argument shot down in flames the moment it was presented? Why has the mainstream scientific community allowed this form of "proof" to go unchallenged? This is one area where the science academies in each country should have shown leadership by requesting that papers and books containing this obviously circular logic be withdrawn. As we have seen even top line journals have accepted almost any material supporting the orthodoxy on climate that they are given, but this form of proof is truly scraping the bottom of the barrel.

There is obviously still a great deal to argue over in all of this but the main purpose of this brief layman's tour of the science is to establish whether there is enough hard, settled science to require that everyone else must pay a steep price to prevent problems, whatever those problems may be. The question is particularly urgent for us as global warming scientists and activists have no intention of paying this price. They consider themselves too valuable. So that leaves you and me to pay the bill. The answer to that key question is a resounding no. But even if the answer was yes, we still face the difficult question of whether there is an economic case for paying a big price to cut emissions now. The answer to that question is also a resounding no, as we shall see.

6
NUMBERS THAT DON'T ADD UP

In 1310, after several centuries of warm summers and relatively stable climate, torrential rains spoiled harvests throughout Europe, turning farmlands recently cleared of forests or reclaimed from swamp into muddy wildernesses. The troubles continued month after month. In central Europe, whole villages were swept away in floods. As archaeologist Brian Fagan says in his previously mentioned book *The Little Ice Age*, few people realised how extensive the rains were until returning pilgrims told of similar misfortune in all parts. To them it seemed like the whole world was in trouble.

The late H. H. Lamb, emeritus professor in the School of Environmental Sciences at the University of East Anglia, says in his book *Climate History and the Modern World* that the change must have appeared "devastatingly sudden", particularly since it came so soon after a period of notably stable climate with mostly warm dry summers from 1284 up to 1311. In that first decade of the 14th century wine growers had the confidence to start new vineyards in England. (Wine cultivation was not finally abandoned in England until the late 16th Century.)

Although the climate had been turning colder for decades before the end of the 12th century out in the North Sea and in Greenland, the Little Ice Age is generally dated from 1300 with the real break from the Medieval Warm Period being the great famine of 1315 to 1321. Up to then the life of the agricultural workers and fishermen who made up the bulk of the working population had been brutally hard, but at least the climate had been stable and harvests (mostly) steady and regular. The resulting famine was made worse by the large populations built up during the good times. (Populations had barely recovered by mid-century when the bubonic plague, or black death, hit Europe. Life was truly hard back then.)

These hard times, interspersed with long periods of better climate, continued up to 1850, the usual date given for the end of the Little Ice Age, with the coldest period thought to be around the end of the 17th century – 1670-1710. As noted by Fagan, Lamb and others the colder climate survives as impressions in history and literature. The novels of Charles Dickens depict particularly cold winters; the winter of 1777-1778 was an ordeal for George Washington's ill-equipped Continental Army and so on.

For a time after the climate variations known as the Medieval Warm Period and the LIA were discovered, some historians tried to prove that almost every historical change was related to climate change. Thus the Mongol invasions were the result of tribesmen finding their usual grazing areas in central Asia drying out in the MWP, and the Vikings took to the seas because of agricultural troubles in Scandinavia. Known as climate determinism this approach has long been abandoned, but climate change of the time had an effect on economies and, as there are sufficient written records to reconstruct a lot of detail, it is possible to study the effects of climate change on European economies. So here is a case study of sorts about how changes in climate affect economies; case studies that may throw light on efforts to estimate economic damage from IPCC climate change forecasts, assuming that those forecasts have any basis in reality. Admittedly it is for a period of cooling rather than warming but we have to take what we can get.

A quick glance at the period does not bode well for those who like big computer models to estimate damages and costs, as a necessary first step in any economic analysis. Different countries were affected in different ways, depending on their political and social systems, cultures and geography. Fagan points out that in England and also the low countries (now Holland and Belgium) farmers experimented and innovated. In England the land-owning gentry tended to take an active interest in their farms and food production. The French aristocracy did not. The farmers as well as agricultural workers in England and the Low Countries adopted the turnip and the potato, both of which are extremely useful in food production. Agricultural workers in France turned up their noses at the vegetables preferring wheat for bread, although wheat does not react well to prolonged rain, and grapes for cash.

For those and other reasons England and Holland fared much better during the Little Ice Age than did France. Conditions were by no means all roses in England as a series of enclosures – the emergence of farms as we know them, as opposed to the medieval manor system – forced a lot of agricultural workers off the land and into the towns to work in the factories of the embryonic industrial revolution. Rather than become involved in the often hot debate between scholars over just why the industrial revolution happened when and where it did, let us just note the agricultural production improved in some areas, despite the colder climate, and was devastated in others.

One frequently cited example of devastation caused by climate change that occurred in the Little Ice Age is that of the Norse settlements in Greenland. These were settled in the tenth century when temperatures were warm and survived to the 15th century. The settlements fared well in the warmer times but when climactic conditions changed the people just did not adapt as they should have. Adopting Inuit methods for hunting seals, for example, would have kept them fed, but they did not and paid the penalty. (See 'The fate of Greenland's Vikings', Dale Mackenzie Brown, *Archaeology*, 28 February 2000.)

Another historical example sometimes cited is that of Easter Island with one writer, Jared Diamond, a professor of geography and physiology at the University of California, claiming that the Easter Islanders managed to destroy their own environment. He says that they cut down all the island's trees to build the monstrous statues that are a feature of the island. (*Collapse: How Societies Choose to Fail or Survive*, Viking Press, 2005.) He says the island's society then degenerated into civil war and even cannibalism and only a remnant existed when the first Europeans arrived. Disease and slave trading wiped out the rest. One piece of evidence for this theory is that the island now has no trees but it is known to have been covered by trees in the past.

Diamond's theories are very strongly disputed by academics in that field, however, notably Benny Peiser an anthropologist at Liverpool John Moores University (and now executive head of the anti-orthodoxy Global Warming Policy Foundation). Peiser says that the Easter Islanders adapted well to their environment and the reason they stopped making the statues is the obvious one, the onslaught of European diseases and mass abductions by slave traders. ('From Genocide to Ecocide, The Rape of Rapa Nui', *Energy & Environment*, 2005.) He says that the islander tales of fights in the distant past relied on by Diamond for his theories have been shown to be confused communal tales of far more recent fights with the slave trader gangs.

Another academic who agrees that European contact is to blame for the collapse of Easter Island's society is Terry L. Hunt, a professor of archaeology at the University of Hawaii, who has conducted several digs on the island. He says archaeological evidence clearly points to the main cause of the demise of the island's tree cover as rats, introduced by the islanders, eating the seeds of the palm trees before they could germinate. Further, at least a part of the island still had trees by the time of first European contact, (*American Scientist*, October 2006).

So much for Diamond. Moving on from these historical examples of varying worth we can also see vast differences in response to environmental circumstances in modern economies. A nation which has resources and is in a warm part of the world, has an advantage over one in the far north, without resources, but a glance at the respective economies of Finland and Nigeria shows that those differences are not the whole story. Finland's 4.5 million people have nothing but a lot of lakes and forests and long, harsh winters. Comparatively little of the land is usable for agriculture. For most of its history the country was under the control of first Sweden, then Russia, and was mainly an agrarian country until the 1950s. Nigeria only shook off its colonial ties in the 1960s but has substantial oil resources, mineral resources of all kinds and productive agricultural lands. Now Finland has the highest per-capita income in Europe, whereas for decades Nigeria has squandered its oil revenues. Although doing much better of late with major increases in its per-capita income, it is still generally thought of as poverty stricken.

There are those who would say that it is not a fair example, as the Finns started with much higher levels of education and are ethnically far more homogeneous (there are lots of different tribes in Nigeria), and there would be some justice in those complaints. Then what about Argentina and Australia? In the late 1940s the two countries were in similar positions with substantial resources and strong agricultural sectors, with Argentina enjoying a higher per-capita income. Sixty years later Australia is an advanced, first world country, and Argentina is classed as an emerging market. Arguably it is one of the better emerging markets having cleaned up its act after the major debt crises of the 1990s (a relative who visited the country recently says that kidnappings are down) but one with a per-capita income that has not changed that much, in real terms, since the 1940s.

A common theme is that the better off countries have generally far less interesting politics and have maintained rule of law, but a full discussion of the reasons for the difference would require a platoon or two of economists. The point is that the outcomes are very different, and that in the case of Finland, and that of Australia noted above, countries can overcame adverse climatic conditions to emerge stronger than before. Argentina and Nigeria had advantages which they wasted. Individuals, industries and whole countries react to circumstances, whether those circumstances are a financial crisis or a shift in climate. Forecasting just how they may react is very difficult, with any forecasts further complicated by shifts in technology. Dire predictions of large sections of Australian farmland being rendered unusable by higher temperatures may be overturned in an instant by the development of, say, a heat resistant form of wheat.

At the beginning of the Little Ice Age a thriving international trade in grain also helped England scrape through the troubles by trading with other areas that managed to keep their agricultural system intact. Similarly, our much more extensive international trading system will smooth out the various penalties and advantages of climate change. If Australia seriously found its agricultural output constrained by climate, then it may be able to import food from, say, Russia which will find more of its land usable thanks to those same shifts in climate. Doesn't that mean the resulting loss of agricultural exports will cripple the country? It is true that one export sector will be affected but the labour and capital employed in the sector would be redirected to some other part of the economy, and earn dollars some other way. The presence or absence of resources is only one part of any country's economic circumstances, and by no means the most important part. Consider the Swiss. Switzerland has a lot of scenery, lots of mountains and lakes and that's just about it. The country's soil has always been poor compared to the rest of Europe. Yet somehow they manage to scrape together a high standard of living. But the Swiss are different, you may say, they are smarter. They are not that different, but they do face a different set of economic circumstances. If they seem smarter that is because they have had to be smarter to overcome disadvantages, and their social and political systems don't get in the way. This is not about more or less government intervention, incidentally. Denmark achieves an excellent economic outcome on very high levels of government intervention.

This is not to dismiss serious climate change, if it ever occurs, but to point out that the world economic system adjusts. Countries that are already rich with adaptable, open economies will probably fare much better, or perhaps not suffer any economic penalty at all. In contrast, badly governed, poor countries may be devastated. Then again, some countries will be able to devastate themselves without any help from climate change.

One industry frequently mentioned as likely to be devastated is that of agriculture, but efforts to determine any likely damage do little more than show just how difficult it is to forecast changes decades into the future. The sector depends on both temperature and rainfall, but what will changes in temperatures mean to rainfall? Will it mean more rainfall or less? As we saw in the historical episode recounted at the beginning of the chapter, a fall in temperatures brought on a deluge, with the same rainfall patterns occurring for several years. Why the dramatic switch? There is no clear answer. Geologists occasionally point out that the "greenhouse" phases of earth's geological past which featured more carbon dioxide and higher temperatures also seemed to be generally wetter. Climate modellers, on the other hand, equate warmer temperatures with less rainfall.

As we have already seen, some short term climate forecasting is possible by watching for the shifts in the oceanic climate cycles such as El Niño and La Niña, but climate scientists only realised recently that rainfall in NSW may be connected to a series of low pressure systems off the coast, and those systems may be connected to the Pacific Decadal Oscillation mentioned earlier in the book. What will an increase in temperature do to those systems? For that matter what will the now expected cooler period of the next few years do to rainfall?

Once modellers get past the problem of forecasting rainfall, they still have to work out what that rainfall, or lack of it, may do to an agriculture sector 30 years and more in the future. To make matters even more complicated for forecasters, farmers faced with problems will look for ways to solve them. This is illustrated by a study of the history of wheat production in America by Paul Rhode of the University of Michigan and Alan Olmstead of the University of California. Rhode and Olmstead recently presented a paper to the American Economic Society looking at how wheat production fared when it moved into parts of the United States where growing temperatures were as much as two to five degrees centigrade higher in the mid-1800s and the late 1900s (*Wall Street Journal*, 3 January 2010). The economists concluded that agricultural production adapted successfully as farmers introduced new strains of wheat that grew well in the new climate.

All this means that economists are basing one set of models on the output of another set – the climate models – which are doubtful to say the least, making assumptions about economic sectors 30 to 100 years down the time track and, on top of all that, are making assumptions about how farmers will react and what new technology they will be able to use. Technological change is, by its nature, unpredictable. Perhaps by then they will be manufacturing meat from raw chemicals? Despite all those difficulties economic models have sometimes yielded useful results five years out, but these forecasts are for many decades in the future! Agriculture is more complicated than most other industry sectors, and certainly more vulnerable to climate change, but that brief discussion should give you some idea of the extraordinary problems involved in any attempt to model the economic effects of climate change. A great deal of economic expertise and a lot of computer power has gone into trying to estimate costs of climate change, but most of it is a waste of time. Stronger words could be used.

Although it is tempting to dismiss all this forecasting as straight lunacy there is some precedent for long term economic forecasting as a "what if" exercise. I have seen financial projections for major projects such as power stations go out several decades, because that is the life of the asset, but not even those making the projections expect them to be taken seriously. Who knows what government regulations on power supply will be in 30 years, or what the electricity grid will be like? Instead the projections are made simply to give the investors a feel for the likely returns of the project and what prices electricity needs to command for the project to stay in the black.

To take the projections out a full century is unusual to say the least, but let us wave away all these niggling doubts and finally get down to the point. Do these theoretical exercises show us that it is worth doing anything costly about emissions now? Mostly the answer is no – a conclusion that would horrify global warmers – for modellers can only make the numbers work if they assume an extreme set of circumstances. One of these is if they assume that the earth really is on the cusp of a tipping point, and that further temperature increases will generate large numbers of devastating hurricanes and floods and so on. Those may cost enough to make the trade off worthwhile. We will get to storms and sea levels in later chapters but there will not be much there to comfort global warmers hoping for apocalypse.

Difficult sceptics also point out that temperatures are known to have been at least a full degree higher in Europe in the Middle Ages with beneficial rather than devastating consequences.

Reality cannot be allowed to interfere with economic models, however, and economists have been known to plug adverse outcomes into their models, only to find that is still not enough. Paying to cut emissions now – assuming we can cut emissions effectively – still just does not add up. To make the numbers say what they want greenhouse economists then have to play fast and loose with the time value of money. In essence they argue that a dollar spent today is worth nearly the same as a dollar in 50 years, for moral reasons. We have a duty to clean up the planet for our grandchildren, or great grandchildren, the argument goes. Therefore the time value of money should be much lower than previously accepted. This argument may be a valid one, although there are plenty who disagree, but voters probably do not realise that they are being asked to spend money on a moral principle rather than an economic one.

The main piece of evidence cited in the argument for spending a lot of dollars to cut emissions is the review conducted by distinguished economist Nicholas Stern for the British Government and using the government's resources entitled *The Economics of Climate Change*. Produced in 2006, close to the Al Gore documentary *An Inconvenient Truth*, it produced quite a stir. The review contends that Global GDP will be permanently reduced by between 5 and 20 per cent by climate change, but the costs of reducing greenhouse gas emissions could be limited to 1 per cent of global GDP each year.

That sounds like a fair trade-off, but the Stern review had a different set of assumptions to other studies, including a review by William Nordhaus, the sterling professor of economics at Yale. Those other reviews mostly conclude that stabilising CO_2 concentrations at 550 parts per million (0.055 per cent of the atmosphere) is not efficient because the costs far outweigh the benefits. The 550ppm figure is, incidentally, what the various models say is needed to limit long-term warming to 2-3 degrees centigrade. As we saw in an earlier chapter the growth in CO_2 concentrations has refused to pay attention to forecasts, so they will not reach that 550ppm figure until very much later than allowed for the Stern or Nordhaus analyses.

The Stern report has since been kicked around quite a bit by economists with Stern and others involved taking the trouble to respond to criticisms and occasionally acknowledging an error, without retracting the main conclusion. As economists are not happy unless they have produced a 600 page report, the debate can quickly become tedious for a non-specialist. However, in January 2008 four members of the Australian Productivity Commission produced what it calls a staff working paper, a short and snappy 114 pages entitled *The Stern Review: an assessment of its methodology*, which outlines the main issues.

The commission's paper says that the Stern report reacts to recent warnings that the earth is warming faster than expected by using higher temperature forecasts of up to 5 degrees centigrade and more, as opposed to other studies which use the mid-range forecasts of around 3 degrees over a century. The report also takes into account the risk of catastrophic outcomes (floods, storms, etc); attempts a more complete coverage of damage assessment (includes deaths from heat and disease), as opposed to earlier studies which only took into account damage to markets; and assumes that poorer countries will continue to be more vulnerable to climate change, although they will be richer by the time climate change takes effect. Lastly, and most controversially of all, it assumes a low discount rate. This is the previously mentioned assumption that a dollar spent today has much the same value, in real terms, as a dollar spent 20 to 30 years from now. We will return to the issue of the discount rate in a moment.

In canvassing these issues, the review paper notes that the literature on the effects of climate change varies widely. Some analyses, for example, emphasise the beneficial effects of the additional CO_2 in the atmosphere. The Stern report draws heavily on studies that give a more pessimistic view of climate change and its effects, and gives little attention to the more optimistic views, but the main problem remains that of discount rates.

The paper says: "Because mitigation incurs costs now for benefits that are expected mainly in the very long-term future, economists use discounting to bring the costs and benefits to a common timeframe. The choice of discount rates is critical. The *Review's* headline conclusion that business-as-usual emissions involve costs and risks that are equivalent to losing 5 to 20 per cent of global GDP, now and forever, is based on discount rates that appear to be around 1.4 per cent per annum. These low rates are the main reason *The Review's* headline estimates of damage costs are so much higher than other studies – many times higher than the estimates of Nordhaus and other prominent economists. Adding one percentage point to the discount rates (that is go up to 2.4 per cent) reduces the damage cost estimates by more than half."

The paper also says the Stern report has attracted both praise and censure but of late has been accepted as establishing climate change as something that can be analysed in cost benefit terms, and taking more account of the "low probability but potentially catastrophic" climate events. It also adds that analysing climate change places "extraordinary strains" on analytical techniques that have been devised for more conventional projects.

One way of looking at discount rates is as real (that is, after inflation) returns on investment. If you had $100 dollars in your hand now what sort of real return would you expect to get by investing it? You would hope for 20 per cent and more but economists assume an average, long term return of 7 per cent less an average inflation figure of, say, 3 per cent, which works out to real investment return of 4 per cent. If you invest $1,000 at 4 per cent without withdrawing any money then over 20 years you have $2,191 and 40 years $4,801 in present day dollars. That does not sound all that exciting but you can see the amount is starting to climb sharply and, as we shall see, the Stern report projects costs over two centuries.

To make life complicated, economists shun the simple example above. Instead they will estimate a future loss and then discount back using the same rate to arrive at the equivalent figure in today's dollars. If we can expect a certain amount of damage in 30 or 100 years time, what amount can we spend now to avoid that future damage and still end up ahead? We need to know the damage to be caused in the far future, which is difficult to say the least, and assume a discount rate. Then we can work backwards to find out how much the damage means to us in the present, in economic terms.

In a letter to the *New York Review of Books* (25 September 2008) written as part of the debate over global warming, William Nordhaus uses a 4 per cent discount rate to calculate that to get to a sum of $US100 million in today's dollars over two centuries, an investor need only invest $US39,204 now. He nominated two centuries as the time frame because, he says, that the Stern report found that most damage resulting from climate change would occur after 2200. The Productivity Commission report cited above says it has been estimated that post-2200 costs account for more than half the growth foregone in Stern's report. Whatever the time-frame the amounts that would have to be foregone now in cutting back on emissions to anything like the levels nominated are enormous. Nordhaus says his economic studies show that balance may be achieved with a price on carbon in the range of $US30 to $US50 a ton. If this was a tax the amount of revenue raised in the US alone would be $US50-80 billion a year, every year. That means climate change would have to cost a great deal in the far future and discount rates would have to be set very low, to make the trade-off in cutting emissions now worthwhile.

Other economists have been more critical of the Stern report. Robert Mendelsohn, a professor in the school of forestry and environmental studies at Yale University in the US, says that one argument against higher discount rates is that they are unfair to future generations. "However, using low discount rates is unfair to every generation; the welfare of future generations will be reduced by low discount rates just as much as current ones." ('A Critique of the Stern Report'; *Regulation*, Winter 2006-2007.) By this he means that the wealth of future generations builds on the wealth of the current generation, but that wealth will be affected by current efforts to restrict emissions.

Among other criticisms of Stern, Mendelsohn points to its treatment of the effects of poorer "low latitude" countries, that is those close to the equator. Instead of recent economic growth rates of 3 per cent, he says the report assumes that income grows at 1.3 per cent, and that populations will grow rapidly. "This combination of assumptions creates in the far future vast billions of poor people living in the low latitudes, the most sensitive region to warming. In contrast, if economic growth was assumed to continue at even 2 per cent and population growth continues to slow, the vulnerable rural poor in the low latitudes would actually shrink in the future." (You will also note that this assumption about economic growth in poor countries is the reverse of the assumption made in estimating CO_2 emissions.)

As noted the Stern report also has its defenders. In the same discussion in the *New York Review of Books* cited above, Dimitri Zenghelis, a senior visiting fellow at the London School of Economics and a co-author of the Stern report, says that an increase in temperatures of several degrees will be devastating and that the costs of action will rise sharply. Particularly destructive events such as large-scale flooding, widespread droughts and intense storms could render some future generations poorer than current generations, wiping out the benefits of economic growth. Stern has also vigorously defended the report that bears his name, writing to the inquiry conducted by Australian economist Ross Garnaut (see below) over the productivity commission's criticism of his report. In the letter he says that if anything the report was optimistic in its estimation of damages from climate change and that his discount rate of 1.4 per cent is in the range accepted by economists albeit at the low end, (*Australian Financial Review*, letters, 7 March 2008).

Readers can look at the evidence cited elsewhere in this book and decide for themselves whether there is sufficient justification for spending billions of dollars on the solutions proposed. Then there is the question of whether any of those costly solutions will be in any way effective, a point to be discussed in later chapters. For the moment one overall conclusion we can draw from this discussion is that the only way to make efforts to cut emissions worthwhile is to make extreme assumptions, and then evaluate the costs of the results ethical rather than an economically.

There is a lot more discussion of the Stern report for those who go hunting for it. One particularly trenchant critic has proved to be Nigel Lawson, a former chancellor of the exchequer in the UK government of Margaret Thatcher, and now a Lord like Stern. Along with a formidable array of other authors he co-authored 'The Stern Review: A Dual Critique' (*World Economics*, October-December 2006) which looks at both the science and the economics and is a modest 67 pages. He has also produced a book *An Appeal To Reason* (Duckworth, Overlook, 2008).

Australia is playing its part in this new industry of climate change analysis. Apart from the Productivity Commission review noted above and lots of individual commentary, the Australian Treasury has produced a report on the costs of reducing emissions. A presentation to the Committee of Economic Development of Australia national forum ('The economic costs of reducing greenhouse emissions: Understanding the Treasury modelling', 11 November 2008) says that those costs will barely be noticeable. This may be true in that the economy will still grow, and we may not realise that it could have grown faster if the IPCC and Stern found something else to occupy their time. Also, although the sums mentioned above for US carbon schemes seem colossal, compared to the multi-trillion turnover in the US economy they are small change. That fact that we may not see the bill, however, is still no reason to pay it. There is every indication that the money will be entirely wasted.

The treasury report says that in the modelled scenarios, "Australia and the world continues to enjoy robust economic growth while making the emission cuts required to reduce the risks of dangerous climate change. Even ambitious reduction goals have limited impact on national global economic growth." Gross world output will grow at 3-3.4 per cent per year in policy scenarios, compared to 3.5 per cent in the reference scenario (where emissions are not cut at all), so that even by 2050 the economy will only be slightly smaller than it would have been otherwise. However, the report also states that it is impossible to foresee the state of Australia and the world in 2050, let alone 2100, and that it does not consider what will happen if Australia decides to go alone in cutting emissions. This is an important point for, as we shall see, there is no hope of a binding, enforceable international agreement on reducing emissions. Therefore much of the economic analysis concerning Australia's efforts at cutting emissions is a complete waste of time.

Then there is the Garnaut report commissioned by the Rudd government. The resulting final version of *The Garnaut Climate Change Review* released in 2008 certainly shows diligence on the part of Ross Garnaut, a professor of economics at the Australian National University in Canberra, and his secretariat. Its 600 plus pages cover most aspects of the debate, pointing to this likely problem and that problem, as well as the occasional advantage to result from climate change, all using projections of rainfall and the like which are, in turn, based on projections of temperatures. Those projections are in turn based on projections of growth in CO_2 emissions which are increasingly being overtaken by real world events.

The conclusions that finally emerge from this long sequence of models makes for grim reading and culminates in a case for doing something now about cutting emissions. As we have already been down this path with Stern we will note a couple of points and move on. In the report's chapter 11, which discusses costs versus benefits, it says, "2005, the atmospheric concentration of greenhouse gases was about 455 parts per million (ppm) of carbon dioxide equivalent (CO-e). In the no mitigation world, under the view of business-as-usual emissions presented in Chapter 3, this would reach 550ppm by 2030, 750 by 2050, 1000 by 2070, and 1600 by 2100."

The 455 ppm figure includes CO_2, methane and the industrial gases and may well be right. But as we have seen methane concentrations are going nowhere and CO_2 concentrations are not paying attention to the forecasts, so the 550ppm figure by 2030 is ridiculous. A more likely figure, assuming there is some growth in current concentration growth rates, is around 500-510ppm, but that should be at the high end of projections. But then who knows what CO_2 concentration increases will do? As for the projections for 2050 let's all have a good laugh and move on. In defence of the Garnaut report compilers they were almost certainly relying on the IPCC projections, which are supposedly the best.

Moving on to chapter six, 'Climate Change Impacts on Australia', the report says that "under the no-mitigation case and through adaptive management, much of Australia could experience an increase in wheat production by 2030. This would involve moving planting times in response to warming and selecting of optimal production cultivars" (the best way to grow crops). "Increases would also result from higher carbon dioxide concentrations. Over time, even with adaptive management, a number of regions would experience substantial declines in wheat yields." In other words things will get better before they get considerably worse. Although the report does forecast the eventual collapse of the entire Murray-Darling basin which produces about 40 per cent of Australia's food, the admission of near-term benefits is interesting, especially considering that climatologists are now saying a few years of cooling will occur before the real warming kicks in. One wonders when the crisis will start.

On a more serious note, in the same chapter the report points to the danger to Australia's natural assets such as the Great Barrier Reef, the rainforests of tropical North Queensland and the Alpine regions of New South Wales, Victoria and Tasmania and, puzzlingly, the deserts of Central Australia, which are already very hot and dry places. The supposed threat to the Great Barrier Reef is dealt with in a different chapter and, like so much else in all the warnings over dangers from industrial gases, vanishes on closer inspection.

In any case, as with the Treasury report, the Garnaut report assumes Australia's carbon reductions will occur along with adequate emission reduction efforts by other countries, and that is just never going to happen.

Another argument occasionally sighted in the economic discussions is that it is better to do something "just in case", and that view is encouraged by the comparatively low cost estimated given by the Australian Treasury modelling cited above. In a paper, *The Economic Science Fiction of Climate Change: A Free-market Perspective on the Stern*

Review and the IPCC (Institute of Economic Affairs, 2008) Graham Dawson, a senior lecturer in economics at the Open University in the UK, says that this argument is best understood in terms of analogy with insurance.

"You insure, not against routine minor mishaps, which are excluded by excess, but against events that are unlikely to happen to you but will be catastrophic if they do. However, this is no actuarial basis for assessing the risks of climate damages because there is no population of individuals whose experience can be used to calculate an average risk."

Dawson also notes that in economic terms the IPCC is a state-owned monopoly trying to impose an industry standard on a science that is only just getting started, and is consuming resources that could be used to develop alternate views of climate. "The IPCC's would-be industry standard is not a science, not a collection of conjectures that have survived rigorous testing, but a politically driven selection from the full range of scientific opinion."

Before leaving the issue of economics and climate change we should look at a major theme of the reports above, that is the vulnerability of poor people to climate change. Poor subsistence farmers living on low-lying areas close to the coast are obviously much more vulnerable to climate change than are the citizens of New York or London. Those richer world citizens of 2020 or 2050 can invest in desalination plants, build barriers to keep out rising sea levels and, if necessary, spend a little more of their much increased income on food. The US, for example, has already produced detailed discussion papers and research on how it can protect both developed areas as well as environmental sensitive areas from the supposed danger of rising sea levels ('State and local government plan for development of most land vulnerable to rising sea level along the US Atlantic coast', *Environmental Research Letters*, 2009). The Netherlands, which obviously has more of an interest in the story of rising sea levels than other countries, is reportedly spending $1.6 billion a year on developing the likes of floating houses and greenhouses, and salt-resistant agriculture. In contrast, poorer countries are much more concerned with surviving day to day.

As we have seen the Stern report is concerned with the welfare of the people of the 23rd century who will, presumably, have the benefits of two centuries of economic growth and innovation. By then the likes of Argentina and Nigeria, will have hopefully learnt from past mistakes and be strong nations able to look after themselves, should there be any climatic changes to tend to. Pessimists will say that is simply time for a fresh set of disasters to kick in, notably the exhaustion of all resources (this is dealt with in part in another chapter).

As that world may be as different from us as Napoleonic Europe is from the present day detailed projections are a waste of time, but the present indicators are that there will be substantially fewer of what current generations call "poor" people (as opposed to relatively poor people, which means those poorer than the rest). This trend, which has gone almost unnoticed and unreported, is largely due to both China and India dumping ideology and adopting market reforms.

All of the above plus the general uncertainty over changes in temperatures and sea levels, plus the general uselessness of attempts to reduce carbon emissions, all add up to one approach – do nothing much. Research into climate change adaption is justifiable, provided that it looks at adapting to colder conditions as well as warmer times, and money should be spent on monitoring climate. But there is nothing to justify drastic action, and never mind the 600 page reports brimming with computer model results, prepared by highly-educated, expensive consultants. These all depend on extreme views of the future which, as this book shows, are not justified. Activists will be horrified by such a bland approach. "It's all too dangerous; it's all happening right away; we must do something now!" Right. The problem is that we are being asked to pay for an expensive cure for a disease that may not exist.

On top of all that the economic modelling cited above assumes that the efforts to mitigate industrial emissions are effective, which they are not. As we shall see, that is another, deeper level of lunacy.

7
CUTTING CARBON

In a well ordered world where matters can be decided without green groups screaming that something must be done, and done their way, governments intending to discourage the production of carbon would place a tax on the substance and call it a tax. Then voters would be in no doubt what it was but would agree with it anyway because they really don't mind paying out a great deal more money for goods and services due to a tax. Right? In a world where lunacy has been allowed to run riot the main solutions proposed for reducing carbon emissions are diplomatic initiatives such as the Kyoto protocol and carbon trading systems.

The only carbon trading system worthy of the name at the time of writing as it is not voluntary is the limited European Union Emissions Trading Scheme (New Zealand may have another limited scheme going by the time this book is published), and that has had such bizarre results that even environmental groups have noticed. One problem is that of emitters being allowed to buy doubtful carbon credits overseas; another is wild swings in the price of carbon. Those swings undermine the main purpose of the market, to provide a stable reference price to allow businesses to plan to reduce carbon. But then neither the ETS nor the Kyoto protocol has had any noticeable effect on carbon emissions in the countries they cover, nor is there any real chance of an effective, worldwide agreement to limit emissions.

China and India, both among the top emitters, have shown

only very slight interest in any agreement, a point confirmed by an international meeting in Copenhagen in December 2009, which was expected to hammer out such an accord. China and India made various declarations which meant nothing, and the only result was a non-binding agreement called the Copenhagen accord. This documents talks about keeping global temperature increases below 2 degrees and requires developed countries that are already party to the Kyoto protocol to increase their commitments to reduce greenhouse gases, and for developing countries to make efforts to reduce emissions. It also talks of developed countries raising many billions of dollars to help developing countries meet those goals.

The agreement was worked out by Brazil, China, India, South Africa and the US, none of which have been noted for their efforts to reduce industrial emissions, is not legally binding and was not even agreed to by many of the countries at the conference. The European Union, the only party at the conference with an ETS in operation, was reportedly not even consulted. Nor was there any useful suggestion about how a binding agreement could be put in place. Preventing any country from, say, repudiating its international debts, going to war, or to stop mass murdering its own citizens is hard enough, without bothering with industrial emissions. This point is amply demonstrated by the sorry history of the Kyoto protocol, examined briefly below. In our non-lunatic world activists who refuse to see this and insist that we must reach an agreement would have been sidelined long ago.

For those who want to reduce emissions rather than grandstand at conferences, a more effective approach may be to set aside some of the multi-billions, that were to be spent on limiting emissions, in a special fund. Properly invested over decades that money could be used to offset the effects of climate change, if and when any change occurred, and to search for technology to clean up emissions at the source. This book does not canvass the various proposed technology solutions, including injecting captured carbon dioxide into deep reservoirs, or turning the gas into pellets or whatever. It also does not look at the possibility of a wholesale switch to nuclear power – a suggestion that just makes activists angrier over a number of supposed problems concerning the storage of nuclear waste. All those issues are worth whole chapters in their own right and, in any case, are not really on the agenda at the moment. However, solutions involving technology would be a lot easier. Granted the technology for reducing emissions is far from effective or workable yet, and nuclear energy has public relations issues, but as solutions they are infinitely easier to push onto both developed and developing countries than emission reduction targets. In any case, the emission reduction targets, as we have seen, are based on what might well be deeply flawed projections.

Activists hate this logic. We must do something NOW! We have to reach an agreement! The earth is approaching various tipping points, the heat is building up in the oceans and so on. But do they want to do something about emissions, or don't they? I can only guess at the mind sets of activists but I suspect that attending international meetings, raising consciousness over the evils of emitting carbon dioxide, and urging everyone to turn their lights out for 24 hours is much more exciting than urging research into technology, particularly as the technology is far harder to understand. Another, and perhaps more persuasive reason, is that urging international action grabs headlines and helps in raising funds. Whatever the reason for this bizarre adherence to an unworkable strategy we should look at existing, patchy efforts to get international agreement on limiting emissions, and at cap and trade systems.

The international agreement, often referred to as the Kyoto Protocol or simply as Kyoto, has many technicalities as well as lots and lots of acronyms. The protocol is to the United Nations Framework Convention on Climate Change (UNFCCC or FCCC). Every now and then the various national signatories have a conference, termed a conference of the parties (COP). So you can have a COP for the UNFCCC, which was the main meeting that occurred at Copenhagen in December 2009. (The COP was one of several meetings.) The protocol set out various targets for cutting emissions, with some variations between the countries, which worked out to an average, overall reduction of 5.2 per cent from total emissions in 1990. The target has to be achieved by the year 2012. That sounds useful but, as we shall see, the targets are largely meaningless. Among other provisions the protocol also allowed for emissions trading and the now notorious clean development mechanism. This mechanism permitted signing countries to meet their targets by buying credits from other countries, including those who are not required to meet emission limits, such as China and India. The credits are generated by hydro-electric dams or solar power stations and so on.

The protocol was initially adopted in 1997 but did not become enforceable until two criteria were met. One was that 55 parties had to agree and the other was that countries responsible for 55 per cent of the world's emissions had to ratify it. These conditions were not finally met, and the protocol made binding, until March 2005 when Russia signed up, reportedly in exchange for admission to the World Trade Organisations as a developing country, and despite strenuous objections from Russian scientists. Two holdouts among the developed countries were the US and Australia, which were also about the only two countries that would have been obliged to do anything to meet the targets. When Australia eventually signed in late 2007, soon after the election of Kevin Rudd's Labor government, by accident or design initially it had a handy way out from what would otherwise have been onerous obligations in the form of an allowance for land clearing and forestry.

This is known as the land-use, land-use change and forestry (LULUCF) adjustment. The UNFCCC compiles various statistics sent in by member countries and these can be found with some searching on the organisation's site (www.unfccc.int). My impression of data is that the each country does a lot of calculations before sending in any figures and "recalculations" are allowed, although the amount of the adjustments are show separately. Figures are also presented both without LULUCF adjustment for forests soaking up CO_2, and with the adjustment.

The latest set of figures for the period 1990 to 2007 lodged in October 2009, just before the Copenhagen convention, shows a mixed bag. Before allowing for forestry changes Turkey is producing 119 per cent more greenhouse gases since 1990, while a host of former

Eastern bloc countries show real reductions. The Russian Federation cut emissions by almost 40 per cent in the period, Germany cut them by 21.3 per cent and little Latvia by 54.7 per cent. But the Eastern bloc reductions are mostly due to the collapse of the Berlin Wall in 1989, as all their economies then collapsed to be slowly rebuilt, with varying degrees of success, in a much more efficient form. Factory production in communist bloc countries was a notoriously messy business. Of the countries that were outside the Eastern Bloc only a handful managed a reduction, of which the most significant is the United Kingdom. But that reduction of 17.3 per cent was achieved largely by the UK's electricity industry switching from coal to gas, taking advantage of an initially plentiful supply from the North Sea. Of late it has had to import more gas for its power stations. A close examination of the results from the few Western countries that achieved reductions will probably reveal similar stories, and there have been no reports of any country cutting emissions by design. The Kyoto protocol has had no effect of any kind.

The adjustment for land use in the latest set of Kyoto figures makes few significant changes but two are worth noting. Latvia receives an enormous boost for land use changes so that it is reported as cutting greenhouse emissions by 478 per cent – it has become a sink for greenhouses gases rather than a source – and Australia becomes a far worse emitter. Before adjusting for LULUCF Australia reported that its emissions had increased 30 per cent. Since its economy has been growing strongly in the period this is hardly surprising. After adjusting for forestry changes its listed as having increased greenhouse gas emissions by 82 per cent. In contrast, the 2006 report on national greenhouse gas inventory data for 1990-2004 also on the UNFCCC site shows Australia's emission rose 25 per cent to 2004, but the forestry adjustment cut that increase back to just 5.2 per cent, or a little above the increase Australia is permitted under Kyoto. Like much else in the global warming debate I am puzzled by the change in the Australian result in the latest figures.

The US increased emissions by 16.8 per cent up to 2007, before adjusting for forestry changes, and 15.8 after the changes, but as it has never ratified the convention it cannot be held to account. The Canadian figures for 2007 are 26.2 per cent before the forestry adjustment and about 46 per cent after, but Canada has already declared that it will not be meeting its obligations under Kyoto. As previously noted Canada cannot be forced to comply. There are penalty provisions in the protocol, detailed on the UNFCCC site, which involves a further reduction in emission level targets for countries that defy the agreement (the first set of targets has already been ignored), and a provision to ban the country from buying credits from other countries. Countries with governments that want to be re-elected, such as Canada and America, or with governments that want to keep a lid on internal dissent like China, are not going to be pushed into making hard choices by these sanctions.

One way for any party to the Kyoto protocol to be seen to be doing something, without actually doing very much apart from impose costs on its taxpayers, is to introduce a cap and trade system. In these systems the amount of carbon dioxide (or industrial gases translated to their carbon dioxide equivalents) major businesses are allowed to emit is set, or capped. If any company wants to give off more CO_2 it can buy permits from other emitters who have gone under their quotas, or be fined. In theory, the penalised emitters will then realise it makes financial sense to clean up their factories, or otherwise invest in emission reductions.

The government which legislated the system into being – the European Parliament, for the EU ETS – also sets the targets and, hopefully, reduces the overall cap on emissions over time. The government will also issue the permits, perhaps for free or for a set fee, or through an auction process. Either way, selling the permits counts as a tax, albeit one that is not readily apparent to the consumers. The bizarre point about this is that the government will be instrumental in setting the underlying price of the carbon market, and in deciding who gets free permits which it can issue at will. Those decisions will almost certainly be made more by reference to politics than to what passes as science in this area. In addition, the government will get an extra pot of money which it can use as it sees fit. Ideally it would give the money back to consumers who will be paying for the ETS through increased prices, but there are always lots of reasons to spend money.

On the surface the European ETS appears to have had some success. *The Annual European Community greenhouse gas inventory 1990-2007 and inventory report 2009* indicates that by 2006 the community was half way to its emissions target. Taking 1990 as the base, and the emissions index at 100, then at 2006 the index was at 90.7. A glance at an analysis of the figures, however, indicates that as with the Kyoto protocol there has been very little progress. The two big movers are the United Kingdom and Germany, but Germany's figures reflect the rebuilding and reorganising of the economy of the old East Germany, as previously noted, and the UK result reflects its electricity industry switching from coal to gas.

Although its record in cutting emissions is doubtful, an immediate and tangible result of the European ETS was to create a market worth many billions of dollars in turnover literally out of thin air. Market tracking company Point Carbon estimates that in 2005, the first year of operation, turnover in the ETS plus several voluntary schemes around the world added up to 7.2 billion Euros. The market has since grown by leaps and bounds. In its annual report on the market *Carbon 2009*, the consultancy says the global carbon trading market was worth 92 billion Euros in 2008. Those are truly staggering sums. No wonder there are plenty of people who believe that carbon must be to blame for something. An American trading system, if one ever gets off the ground, would raise those sums into the stratosphere.

The market is not without its risk, however, with the first few years of trading characterised by wild swings in the price of carbon. For a period in 2007, partly thanks to various governments issuing lots of free emission permits, the price of carbon fell to zero, and that was just the start. A UK group The Taxpayers Alliance produced a report in late 2009 attacking the ETS saying that the emissions price has rapidly fallen by one third or more several times since trading started in 2005, and that it amounts to a regressive tax. That is, its burden falls more on the old and poor who spend proportionately more of their budget on electricity. The electricity industry is one of the main users of the European ETS. A point to note here for reference in the chapter on wind power is that ETS price on carbon emissions was around 15 Euros a tonne for most of 2009. In the early part of the scheme in 2007 it managed to get above 30 Euros.

There are many more flaws in ETS schemes in general, including the tentative US proposal and the Australian Carbon Pollution Reduction Scheme which was rejected by the Australian Senate in late 2009. This rejection featured a major Liberal party room revolt against Federal leader Malcolm Turnbull, who favoured allowing the Labor government's ETS legislation through the Senate. Tony Abbott, a minister in the former Howard government, narrowly defeated Turnbull in the resulting ballot. One of his first decisions was to block the legislation. The Labor government has since vowed to reintroduce it, but this book does not look at the politics of any of the ETS proposals, or at many of the flaws in such schemes. Now that any possible international agreement on emissions has receded into the far future, there is no point to an Australian or even an American emissions trading scheme.

There is also very little point to the European scheme with any justification undermined by one aspect of it that truly deserves the label of lunacy, that of the Clean Development Mechanism (CDM). When negotiators were framing the Kyoto protocol they added in a couple of adjustment mechanisms so that individual countries could meet targets without too much trouble. One of these was the CDM which involves emitters being able to buy offsets or permits of Certified Emission Reductions from overseas. (So the CDM involves trade in CERs. To avoid acronym overload syndrome or AOS, instead of CERs we will talk about offsets or carbon offsets.)

CDM was started with the best of intentions – with a name like that, how could it be wrong – and was initially considered experimental, but quickly grew far beyond the original intentions. The idea is that companies or countries short on emission certificates can buy carbon offsets from clean and green projects in developing countries such as dams in China, natural gas plants in India and wind farms in Indonesia. This presents various problems that have been pointed to time and again by a range of commentators with the latest at the time of writing being the UK green group Friends of the Earth. Just before the Copenhagen conference in late 2009 FotE produced a report entitled *A Dangerous Distraction*, which has some harsh comments about the CDM.

The report says one of the big problems is of proving "additionality". That is proving that buying carbon credit from a project is a genuine incentive for that project, which is commonly a dam in China, rather than an additional profit for a project that would have gone ahead anyway. Then there is the problem of deciding just how much carbon would be saved by the project. How much demand would there be for the electricity produced by the dam and could that demand have been met in other ways?

In late 2007, a US non-government organisation International Rivers issued a report 'Failed Mechanism: How the CDM is subsidizing hydro developers and harming the Kyoto Protocol' saying that by the end of 2007, hydro projects were expected to make up 15 per cent of CERs. The bulk of those dam projects are in China, the world's most prolific dam builder, with an astonishing 25,000 plus major dams. There is no evidence that the extra money the dam developers earned by selling carbon offsets made any difference to whether the project went ahead or not. The dam developers just keep on rolling them out. As of late 2007 about 650 projects had gained, or were set to gain, carbon credits from the CDM for sale. International Rivers also keeps an up to date list of projects which shows that in early December 2009 1328 projects were registered or were seeking registration with the CDM. Of those 65 per cent were in China. Of the projects considered "large", that is of more than 15 megawatt capacity, 71.4 per cent were in China.

Another major and almost intractable problem in assessing these projects is that even if it can be shown that less carbon is being used in one project in one part of a foreign economy, that may still not add up to an overall saving. Investment in a dam or a wind farm may have diverted funds needed to make an old coal-fired electricity plant more efficient. In fact, the complications are endless. A dam in China may be useful but there is unlikely to be any formal assessment of what the dams may do to local ecology, or even the people they displace. Also, the dam may simply be substituting for a coal fired plant that has not been built, so it should not count as a reduction in emissions.

Yet another report released in April 2008 illustrating just how the whole CDM process has become bent out of shape is *A Realistic Policy on International Carbon Offsets* written by two American academics Michael W. Wara and David G Victor. Wara is a research fellow at Stanford Law School and Victor is director of the Program on Energy and Sustainable Development at Stanford University. In the report Wara and Victor say that the CDM executive board which evaluates these projects is massively understaffed and relies on third party verifiers (a form of auditor) to check the claims made by project developers. But those verifiers are paid by the project developers and so have strong incentives to approve the projects they check.

A major part of the problem is that proper checks, including assessing how the project fits with state energy policy, cost a lot of money and the bill has to be paid by someone. The two academics say the CDM executive board "faces a financial limit on the costs it can reasonably impose on individual offset projects. In order to remain viable, relatively small carbon offset projects cannot afford the cost and uncertainty that would accompany truly extensive scrutiny. Indeed there is strong pressure from CDM investors to limit such transaction costs and speed up approval".

They say that the growth in market has been "truly extraordinary". In 2007 the CDM market totalled 12 billion Euros, more than triple the previous year's figures. The European Union's ETS is the biggest buyer of these offsets with Wara and Victor estimating that of the reductions in emissions required in the European scheme will be largely met by buying offsets from overseas. A major part of the reason for this, they say, is simply that it's cheaper for emitters in the EU to buy the credits from overseas than work out how to save emissions at home. The FotE report cited above also says the use of CDM credits means that emitters in the EU have far less incentive to invest in ways to reduce emissions.

The CDM authorities have been making an effort to fix the problem. The EU has committed to reduce its emissions by 20 per cent by 2020, but now there are restrictions in place so that CDM offsets can only be used for 10 per cent or half of that reduction. The Point Carbon report cited above says that the CDM board has been taking a tougher stance on CDM approvals. Other reports say that new dam proposals are now being excluded. Wara and Victor cited above say that there is little oversight of the integrity of the verification process and no record that any verifiers have been punished for misconduct, but that has changed since their paper was written. The board has reportedly disciplined verifiers and has greatly increased the staff responsible for checking verification reports.

Before that the board had taken action on the very worst effects of the CDM scheme, that of awarding credits for the reduction of trifluromethane or HFC-23. Wara and Victor say that HCF-23 is a potent greenhouse gas produced as a by-product of the manufacture of another gas HCFC-22, used in air conditioning units and as a feedstock for the manufacture of high-performance plastics. Capturing and destroying HFC-23 is cheap and easy to do, and Western manufacturers have long done so voluntarily. Manufacturers in the developed world, however, simply vented the gas. The advent of CDM overnight transformed those manufacturers into ventures that manufactured carbon credits with a sideline in making refrigerant gases. They were producing the refrigerant just to destroy the waste gas and claim the credit. They say this "perverse incentive" has been limited by the CDM executive board but not eliminated.

Ross Garnaut, a professor of economics at the Australian National University referred to in other chapters, also produced a report on the CDM problem as part of his work on the Australian emission trading scheme. The *Emissions Trading Discussion Paper* released in March 2008 points to the problem of additionality, and of the difficulties involved in linking the Australian scheme to overseas ETSs. This warning was not heeded. The Australian scheme if and when it gets off the ground, will have access to overseas offsets, although consultants are divided on just how much Australian emitters will use offshore credits.

Garnaut's paper was, as we have seen, just one of a long line of reports and papers pointing to the problems of CDMs along with suggestions for reform. The Friends of the Earth want to abolish the scheme altogether as simply a transfer of money to no purpose. Wara and Victor want to reform it, with one suggestion being to deal with the case of refrigerants outside such a system with a separately targeted fund. There has also been talk of extending the CDM concept to allow credits for changes in forestry management. Then there are those who suggest that perhaps we should dump the idea of a trading system and simply impose a tax on carbon emissions with that tax adjusted so as not to impose an additional burden on exporters.

Another flaw that has provoked much discussion is the problem of individual governments being permitted to issue permits for free. They are allocated a certain number of permits under the ETS rules, but have some discretion in what to charge for them. There have been reports of this manufacturer and that European manufacturer threatening to go elsewhere, or whole sectors claiming they will have to shut down, unless they get more free permits. There are plans to centralise the issuing of permits in the European ETS, but this will not happen for several years.

To sum up, these trading systems are proving immensely difficult and complicated, and require considerable political will to make them work properly. That political will is unlikely to be forthcoming. However, one result of having a carbon market is to create a whole class of carbon traders, brokers, advisors and consultants of all stripes earning comfortable livings from it. All of those people would have a vested interest in ensuring that the market system continues, and that means ensuring that no one threatens the orthodoxy that industrial gases are contributing to climate change. Suggestions that the market may be a highly inefficient means of reducing carbon are given a similarly short shrift.

With the carbon market representing a lot of political trouble for little return it is not surprising that of the big four emitters – China, India, the US and the EU – which collectively will emit two-thirds of all carbon dioxide from industrial activities between now and 2050, only the EU has a carbon trading scheme. China and India are developing rapidly but have never been a part of Kyoto and have shown only slight interest in the Western obsession with carbon. Before the Copenhagen meeting at the end of 2009, China's chief contribution to the debate was to demand money in return for cutting their emissions. At the meeting both China and India promised a reduction in carbon intensity, that is they would cut the carbon emitted relative to the growth of the economy. Emissions would still increase, just not as fast. China promised to cut carbon intensity by more than 40 per cent by 2020, India by 20 per cent. For its part the US has already proposed absolute cuts of 17 per cent while the EU has pledged to cut emissions by 20 per cent by 2020. All of these cuts are promised for different base years but there is no real need to worry about the fine details, as they are all meaningless. There is no means of enforcing those promises and no mechanism for checking the results beyond trusting dodgy government statistics. On top of all that the situation is even more confused by the projections for CO_2 concentrations being almost certainly wrong.

President Obama will make an effort to be seen to be doing something about Kyoto promises but has to get legislation through the American congress and senate, and that seems doubtful. There are very powerful forces in US politics that are far from convinced by the carbon dogma. For the EU to gain cuts of 20 per cent would require the union to display considerably more political will than the easy ride it has had of late. In any case, as we have seen, thanks to the CDM the actual cut will be closer to 10 per cent not 20. India may make an effort, and will certainly invest in solar energy if only to keep the Western powers quiet, but whether its efforts succeed is another question, particularly as there are unlikely to be any independent checks. In any case, the pledge only relates to energy intensity and not to overall reductions.

Then there is China. Ordinary pollution, including poison in rivers and particulate air pollution (smog) in cities, is notoriously bad in that country, and Westerners are fussing over carbon dioxide? As part of that fast-track development, and has been pointed out time and again, China is building vast capacity in coal-fired electricity plants besides the dams we have already discussed. In the paper cited above, Wara and Victor say that China is building about 100 gigawatts of new power plants each year – an annual addition far more than the total Australian generating capacity and about one-tenth the total US capacity.

Perhaps the Chinese central government would make an effort to meet the energy-intensity targets, and it certainly will invest in solar plants, with the help of contributions from the CDM scheme, but the relevant statistics are usually whatever the government thinks they should be at the time. Binding China to any agreement at all is even harder than it is for most other countries. In China, for example, a written, signed contract is not seen as binding, but merely what the parties think the deal should be at that stage of the negotiating process. At best, party officials under pressure from the central government may fudge the books to make it look as if something has been done. Worse, those figures may be taken at face value, as have the figures on economic growth.

Don't believe this? Read *Mr China* by Tim Clissold (Random House, 2004), *China and the West in the 21st Century* by Will Hutton (Little Brown, 2006), *The Asian Insider* by Michael Backman (Palgrave Macmillan, 2004), or *One Billion Customers* by James McGregor (Nicholas Brealey Publishing, 2005). Aside from the mere technical details of enforcing an agreement, the Chinese government is unlikely to curtail or reduce growth over real fear that any economic slowdown may create serious civil unrest – unrest which may raise unwelcome questions about its legitimacy.

As previously mentioned, rather than chase the impossible dream of international limits on emissions it would be better to put a fraction of the money that was to be spent limiting emissions into a fund to help future generations fight climate change, if and when it occurs, or to urge further research into carbon saving technologies. But our noisy activist sector considers those options to be completely unsatisfactory. They are not exciting, dramatic solutions which feature countdowns, or one that appeals to the innate beliefs of that sector – namely that human industrial activity must be to blame for something. Perhaps most importantly of all, it does not help non-government organisations raise funds.

Now we come to the really tricky question of whether any of the proposals acceptable to activists as ways of reducing emissions, notably alternative energy projects, actually reduce emissions. This is yet another level of lunacy.

8
BLOWING IN THE WIND

One form of climate change lunacy which has been adopted worldwide with very little opposition, is that of wind energy. But that near universal adoption, at least among Western countries, has little to do with its effectiveness. In fact, about all wind systems seem to do is to jack up electricity bills without making a dent in emissions. Rather than reduce emissions, the main use of wind generators now spreading through England, Denmark, Spain, Germany, America and Australia and most other advanced countries you can name, is as political symbols.

Renewable energy projects, of which wind generators are by far the largest and most visible component, are there to make voters think that their government is doing something about emissions. Further, they are a gesture the government can make without having to raise any taxes or spend very much money. The Australian scheme is typical of this approach, with the government pushing legislation through Parliament in late 2009 which required that 20 per cent of national electricity consumption must come from renewable energy. Consumers will pay for the government's foresight and green energy policy through increased electricity bills.

This wide-spread adoption of wind energy, despite well-documented problems, has been helped by the lack of any organised opposition. Energy companies do not care as the use of wind hardly affects the number of power stations that have to be built or the energy that they use, and it is not the job of electricity network managers or energy supply companies to force the government and consumers to see sense, particularly when they can pass the costs of this lunacy onto the consumer. There is some anecdotal evidence that energy industry executives and grid managers are, in any case, reluctant to point out the shortcomings of wind power, for fear of the reaction from the noisy activist sector.

At first glance there does not appear to be anything wrong. A wind farm, say one capable of generating 100 megawatts, will supply it's energy to the electricity grid for purchase by a large retailer (the companies that distribute electricity to homes) or a major user that buys directly. The fossil fuel generators will reduce what they supply by the corresponding amount, saving 100 megawatts in emissions. If they keep it up for an hour then its 100 megawatt hours, the common unit for measuring electricity in bulk. Despite the wind being free, electricity from wind farms is much more expensive then power from coal or gas plants. A conventional power plant will operate from 40 to 60 years, but a lot of wind towers are required to substitute for one power plant and those towers have to be replaced perhaps every 15 to 20 years. But that extra cost buys a reduction in emissions, so everyone should be happy. Right?

Sadly, it is not that simple. The supply and demand of electricity on a power grid has to be balanced at all times. If demand goes up, as it does in the evening when workers come home and start cooking dinner or switch on the television, grid managers have to put more power onto the grid. That usually means connecting generators to it, or directing generators already connected to increase their power. If they don't then voltages fall throughout the grid and there are brown outs in part of it.

The same thing happens in reverse when consumers go to bed and demand falls. If the resulting unused power is left on the grid it would change the voltage and frequency of the supply, which would greatly upset electrical appliances. So generators have to be instructed to reduce power or be taken off the grid altogether. At all times the grid managers have to balance supply and demand on the grid, as well as cope with unexpected peaks in demand such as exceptionally hot days when everyone turns on the air conditioner. With one exception to be explored later electricity cannot be stored on power station scale. It is produced at the same time as it is consumed with nothing left over, no refunds, no returns, no exchanges.

To keep the grid balanced and consumers happy, grid managers will use several different types of generators. The backbone of the supply is the big coal and nuclear powered plants which operate almost continuously to supply the majority of the power – the base load. The output of those stations changes only slightly over time and they are not shut down lightly, as they can take a whole day to restart. Brown coal and nuclear plants in particular cannot vary their output very much at all. If they do it is by venting steam, resulting in the white clouds often seen coming from power stations. The clouds are not pollution but steam. Black coal power stations can vary their output but not easily and not very quickly.

Then there are the open and closed cycle gas turbines. Closed cycle turbines are bigger and more efficient than their open cycle brethren, as they recycle part of their heat, but are more difficult to power up and down when demand changes. Open cycle turbines are much less efficient, but cheaper to build and can be powered up and down more easily, so they are the most common "load following" generators. Then there is hydroelectricity which is the most responsive – that is the easiest to adjust and switch on and off – and so is used for the peaks in demand. Hydroelectricity also counts as a renewable but does not cause anything like the problems of wind generators, as hydro turbines can be switched on or off at the grid manager's direction. If all the 20 per cent renewable target came from hydroelectricity there would be no problem. As matters stand about 4 per cent of power in Australia comes from hydro-electricity, but there is no prospect of that proportion increasing.

Unlike hydroelectricity and all the other types of turbines discussed, wind energy cannot be turned on and off at the request of the grid managers. It turns itself on and off and, as the grid managers often have to accept the power to meet government requirements, the rest of the grid has to be organised around this intermittent source. Wind farms greatly complicate the 24 hour balancing act of managing grids. Let us suppose our wind farm suddenly starts producing electricity at mid-morning, past the peak of commuters taking showers, making breakfast and getting themselves to work. A network which has not been redesigned to cope with this influx will only have the big coal generators operating at that time of day. To compensate for the extra supply of energy the grid operator then has to partially "deload" the coal plant – perhaps by switching off one of its generators or making it vent steam but keeping the plant turning at the same speed – or take it off the grid entirely so that it is spinning away but not producing anything.

Those big plants are designed to operate efficiently at one speed and load. If conditions change that efficiency plummets and emissions increase. Depending on circumstances the extra one hundred megawatts of clean wind energy may even boost network emissions. It will not be that bad, of course. The grid will be set up so that in place of one of the big, efficient base load plants there will be several load following open cycle gas turbines which can be powered down as wind comes onto the network, or taken off the grid with much less trouble. Open cycle gas turbines are generally less efficient and emit more gases than a properly-sized closed cycle gas plant, and their performance suffers if the operators have to keep powering them up and down because the wind keeps on changing. The supposed 14 per cent reduction in emissions (20 per cent less 4 per cent for hydro) is then reduced by a few percentage points, and there are more problems to come.

What happens if the wind stops blowing or, more likely, suddenly falls by one-third? The grid operators will have to make up the shortfall from somewhere. Open cycle turbines are flexible but not that flexible which means that some generators have to be kept operating at all times, off the grid, ready to hooked up when there is a sudden change. Grid operators always keep operating plants off the grid ready to be hooked up in an emergency, such as one of the conventional power plants failing unexpectedly. Just how much of this "spinning reserve" is kept off line, generating emissions but not producing power depends on the risks involved. What is the risk that an operating plant will have an unexpected "outage"? One approach is for the operators to keep a spinning reserve equivalent to the largest operating plant on the grid but for large networks with many power stations, the operators will work out the risk and calculate what reserve has to be kept offline to be connected in seconds, five minutes and 15 minutes and so on. They also have procedures for calculating how much "load following" capacity they require to meet unexpected changes in demand.

In various overseas reports consultants point out that conventional plants also have their variations and may only be operating, say, 85 per cent of the time. True, but most of those outages are planned, such as for maintenance, or are requested by the grid operator. Wind comes and goes, so the risk of an unplanned outage is very much greater. That means the operators have to keep more spinning reserve to hand – a reserve generating emissions which the wind farms are supposed to be preventing. Our supposed savings from using wind have been reduced again but by how much? As always the devil is in the considerable detail, and we will return to the point of spinning reserves.

A peculiarity of wind farms is that they stop operating when the wind is too low (wind speed must pass a threshold point) and when the wind is too strong. For various reasons wind towers cannot be built strong enough to withstand winds from storms and so the blades are deliberately set to "feather" (the angle of the blade changes) to avoid the full force of very high winds. So our wind farm in the example above will sometimes produce nothing, at other times will produce 1,000 megawatts and at other times a value somewhere in-between. One state in a position to pass judgement on the operational difficulties of wind farms is South Australia.

In 2009 the Essential Services Commission of South Australia produced a draft report on licence conditions for wind generators which notes that the state has 3,641 megawatts of installed conventional generators (fuelled by coal, natural gas and distillate) and 739 MW worth of installed wind towers, with another 128 MW under construction. By the end of 2010, says the commission, the state will have more than 1,000 MW of wind generation capacity, and that capacity should be seen in the context of the summer peak demand in South Australia for 2008-09 which was 3,450 MW. This makes SA truly the wind centre of Australia, with the commission noting that only Denmark has a higher contribution from wind relative to customer sales.

The commission decided to continue the licensing of wind farms but with tougher conditions so that they can more easily withstand problems such as power spikes or sudden drop-outs called "fault ride-throughs". However, the draft report's conclusion expresses concern over the "long term safety and reliability" of the electricity system of the state system with so much wind power to hand, unless precautions are taken. When the report was released commission chairman Pat Walsh told the *Adelaide Advertiser* (26 June 2009), that on average wind farms only delivered about one third of their installed capacity, and could only be relied on for about 10 per cent of their capacity to meet peak demand. Mr Walsh declined to return my calls as a journalist and a commission spokeswomen referred further questions to the National Energy Market Management Company, which is now the Australian Market Energy Operator. The new entity, which operates the connected grids for Eastern Australia and South Australia, also includes a gas market.

Walsh's estimate of wind farm output averaging one-third capacity is broadly in line with, although at the higher end of, overseas reports. Reports for wind farms in England, a windy place by all accounts, generally puts the average output at 25-30 per cent of capacity. A typical capacity factor for a conventional plant is above 85 per cent but, as we have seen, the 15 per cent the plant is offline is largely scheduled and planned for. Now that we have a capacity figure we can calculate the number of wind towers required to replace a conventional power plant. A major coal plant typically has a capacity of around 500 megawatts. A single wind turbine may have a rated capacity of 2 megawatts. If they are all going at full stretch then 250 would be enough to replace the coal plant. As we have seen the average output is a third of rated capacity so we would require 750, all well spread out so as not to interfere with each other's wind flow. Only isolated windy spots will do for such an installation, but even that 750 figure is not a true comparison as wind is so unreliable compared to conventional power plants. Wind towers now coming into use in England are rated at much more than 2 megawatts but are gigantic structures (see below) and individually they will require a lot more space.

Although wind matters a lot in South Australia, nationally it is still only a small proportion of supply. In mid-2009 renewable energy sources accounted for about six per cent of supply with hydroelectricity taking four per cent, leaving around 2 per cent for the rest which is mainly wind. Absorbing such a small amount of wind energy is comparatively easy on a large grid. The real problems come when wind energy equivalent to perhaps 5 to 10 per cent of total generating capacity is connected. To have 10 per cent of total capacity suddenly appear on the grid or disappear from it, can be startling so that the grid operator has to consider how much spinning reserve and/or load following capacity is required.

There will be many factors to consider. Electricity grids are, by their nature, run very conservatively and designed to cope with worst case scenarios. Grid managers cannot simply brush aside the worst case involving wind generation – a hot, calm day – on the assurance that it does not happen all that often, or that there will always be wind blowing somewhere in eastern Australia (the west Australian grid is separate). The Clean Energy Council of Australia, the main wind group, points out that Australia is a very large place and that if the wind farms are spread out far enough there will always be one operating somewhere. Electricity industry executives, however, counter this by pointing out that generators have a transmission radius. They can transmit electricity only so far over power lines. A wind farm in Queensland cannot fill the gap if the wind in South Australia stops blowing.

While we are on the subject of transmission distances, another problem with wind is that the most promising sites are usually in remote locations, such as on the coast or outside Broken Hill, well away from the major population centres. Those remote locations require additional transmission towers which have to be built to take the maximum output of the farms, although most of the time they will be generating much less, depending on the wind. It is akin to building super highways to handle traffic that, for most of the time, is at the volume found on a suburban road. This requirement further reduces our supposed saving in emissions, but no one can say by how much.

Australia is only starting the process of working out how to put 20 per cent renewables on the grid and the additional costs of connecting to wind farms are unknown. A 2007 report *Intermittency Analysis Project: Final Report* produced by the California Energy Commission, which is aiming for 33 per cent penetration by 2020, estimates that it will need to spend US$5.7 billion on transmission towers and another US$655 million on new and improved transformers to meet the wind target. The Californian economy is obviously very much larger than that of Australia and its target is much higher but those figures give a taste of what may have to happen in the Australian network.

Besides spending money on new transmission lines and on lots more load following generators, the industry will have to spend big on generators of all types. Until grid managers know much more about wind patterns, and the output of wind generators, they will have to design for the worst case scenario of hot days with no wind. Reserve requirements will also be on the cautious side. In other words the network will have to be designed to have sufficient capacity if the wind farms were not there at all, and still have reserves for emergencies. When there is plenty of wind power on the grid, managers will still plan to have a lot of spinning reserve in hand, just in case. Planning grids so that 16 per cent worth of generation capacity will not be operating at crucial times, and operating at full capacity when it is not wanted, is odd to say the least but that is what the government has decided will happen with no protest from voters.

How often do hot, calm days occur in Australia? The issue remains almost unstudied, as far as this writer knows, but the reverse of cold, calm days is quite common in Europe – depending on who you speak to. One German company with a lot of wind in its generation portfolio and prepared to look at this energy source realistically is E.ON Netz GmbH. In its *Wind Report 2005* the company notes that "both cold wintry periods and periods of summer heat are attributable to stable high pressure weather systems. Low wind levels are logically symptomatic of such high pressure weather systems. This means that in those periods the contribution made by wind energy to meeting electricity consumption demand is corresponding low." A more recent report by the UK Energy Research Council, 'The Costs and Impacts of Intermittency' part funded by the pro-green Carbon Trust acknowledges the "low wind cold snap scenario" but says that long term meteorological records indicate that they rarely affect all of Britain simultaneously. In contrast, in a 2008 article in the journal *Energy Policy* entitled 'Will British weather provide reliable electricity?' gas industry consulting engineer Jim Oswald and co-authors also looked at the meteorological record and found that calm conditions across all of the UK are regular events. Further they found that the calm conditions often extended across a large part of Europe.

The usual reaction of wind advocates is to dismiss the Energy Policy article and the E.ON report as part of a big energy conspiracy (E.ON also runs conventional power plants) and accept the UK ERC report, ignoring the quibble that is has been part funded by the Carbon Trust. But grid managers must still take the cautious approach and prepare for no-wind, high demand scenarios and plan for additional reserves. They can arrange for a spread of wind farms to smooth out variations in wind energy, and so reduce the previously-mentioned reserve requirements. Network operators around the world are also working on wind forecasting systems in order to better judge likely wind energy supply at any given hour or day, and that may also help reduce the reserve requirements.

So when all is said and done how much is likely to be saved in emissions and at what cost? Anyone who trawls through the endless overseas reports on this issue will eventually realise that the numbers can be arranged to say almost anything, particularly if the authors are investigating the performance of wind farm networks that have not yet been built. Spinning reserve requirements can be calculated in different ways, and different assumptions can be made about how electricity will be traded across network connections. There is no agreement on how to report the results of these various studies, and no standardised terminology.

There are any number of studies from environmentalists to indicate that the use of wind may not cost all that much. A paper by a Finish researcher Hannele Holttinen, 'Estimating the impacts of wind power on power systems – summary of IEA Wind collaboration,' (*Environmental Research Letters*, 2008), states that for 20 per cent penetration (contribution) of wind, various studies show that the increase in the wholesale cost of electricity will be 10 per cent or less. Another report, *20% Wind Energy by 2030* from the US Department of Energy released in 2008 also looked at a number of surveys estimating costs and came up with a similar figure. A 10 per cent increase may sound bearable, but that increase comes from just 20 per cent of the energy supply, which implies that wind power in general is up to 50 per cent more expensive than fossil fuel plants, after taking reserve requirements into account. A closer look shows that these are mostly academic studies and, crucially, the analysis does not give any overall figure for the amounts paid per tonne of carbon emissions avoided.

Another theoretical exercise, a 2004 study by the Royal Academy of Engineering, *The Costs of Generating Electricity*, came up with considerably more pessimistic results. Engineers are usually pessimistic about new technology, it's what they do, but the academy is less likely to have a political agenda. The report estimates that electricity from an onshore wind farm is 68 per cent more expensive than electricity produced from the closed cycle gas turbine, without taking the cost of additional spinning reserve into account, and almost 50 per cent more expensive than coal powered plants. The engineers calculated that reserve requirements increase wind farm costs by 47 per cent, which means that overall the costs are double that of a conventional power project and most of the emissions savings would be wiped out. Offshore wind farms produce even more expensive electricity (England has a lot of offshore wind farms), and wave energy plants are more expensive again. A more recent study by an experienced power engineer, Peter Lang, as part of a submission by the anti-green Carbon Sense Coalition to the draft version of the recently passed renewables legislation, came up with similar figures.

The 2005 E.On report cited above says that because wind is so variable "traditional power stations with capacities equal to 90 per cent of the installed wind power capacity must be permanently online in order to guarantee power supply at all times". As E.On is about the only company with substantial wind portfolio which has said anything about the spinning reserve requirements, its statements should trump any number of studies. There are a few caveats. The 90 per cent figure, which would completely eliminate any savings in emissions, includes normal reserve requirements whatever they may be (the report doesn't say). Also, at the time of the report cited above, for various reasons, the company could not access wind farms across Germany, just its own which are concentrated on the Baltic Coast near Denmark. The wind off the Baltic is extremely variable.

None of the company's subsequent wind reports are readily available but a pamphlet produced by the company entitled *Carbon Cost and Consequence for the UK market*, when the UK also decided to adopt a target of 15 per cent of electricity to be generated by renewables by 2020, talks about containing electricity price increases to "acceptable levels". In contrast, the UK ERC report cited above dives into calculations in the same area to estimate that backup requirements for wind intermittency adds just 14 to 16 per cent to the cost of wind energy. Note that the report talks about additional costs rather than reserve requirements, or emission savings. Wind advocates like the ERC figure and reject the E.On report as dated.

One problem in deciding how this all applies to Australia is the major differences in circumstances. The areas involved are larger, the populations smaller and there are no way to export electricity, which has proved to be all important in the case of Denmark. As noted above, Denmark has a high penetration of wind and has opted for an even higher penetration, again more for political reasons than due to any confidence in the technology, and in the teeth of considerable evidence that wind energy generated in Denmark is of limited benefit.

A report *Wind Energy The case of Denmark* produced by a prominent conservative think tank, the Center for Politiske Studier (Centre for Political Studies or CEPOS) in September 2009 estimates that Denmark produces the equivalent of about 19 per cent of its electricity demand with wind turbines, but only about half of that is used locally. The rest is exported to Sweden, Norway and Germany through connections to the national grids, and those countries use the energy to pump water uphill into their major hydroelectric dams. Hydroelectric projects are the one viable exception to the problem of storing power on an appreciable scale, as water pumped up into a dam can later be released to power turbines. As well as having lots of dams, those national grids are much larger than the Danish network and so can absorb the additional power easily.

Unfortunately for Denmark, however, not only is the electricity exported at knock down prices - in fact occasionally for free (there is talk of paying the other countries to take the electricity) - it has to be imported at high prices at times of strong supply. This situation and the problem of balancing grids has not been helped by a Danish legal requirement that wind power must be accepted onto the grid when it is produced. Although wind power has saved some emissions in Denmark (the report does not consider reserve requirements), the report estimates that the cost per tonne of CO_2 is 87 Euros or $US124, or more than six times the price of carbon on the ETS at the time of the report. Danish domestic electricity prices are among the highest in the European Union, although this is not strictly the result of wind power. Commercial and industrial prices are deliberately kept down to make industry competitive with the rest of Europe. There is worse to come. The report says that the 50 per cent requirement will require a complete re-engineering of network.

A closer look at results from another major user of wind power as well as photovoltaics in Europe, also indicates that about the only notable result of wind power has been very expensive electricity. A report by the Rheinisch-Westfalisches Institut für Wirtschaftsforschung (a leading economic research institute based in Essen) issued in October 2009 says that extensive subsidies for wind power mean that the country has the second largest installed wind capacity in the world, behind the US. Subsidies for photovolatics (silicon wafers that generate electricity when the sun shines on them), have also resulted in the country having the highest installed base of PV in the world, ahead of Spain.

Despite having a lot of wind turbines and retail distributors of power, paying wind farms three times the going wholesale rate for power (through the mechanism of "feed-in tariffs"), only about 6.3 per cent of total power consumption is supplied by wind. As a result, the report estimates, wind subsidies account for 7.5 per cent of consumer electricity prices. The contribution from photovoltaics is a negligible 0.6 per cent, despite utilities paying eight times the going rate for electricity from those projects. Those estimates for actual grid operations indicate that of the studies cited above the Royal Engineers were much closer to reality, as far as costs are concerned, and the others estimating a 10 per cent increase for a 20 per cent penetration wildly optimistic. It gets worse. Further into the report, the institute estimates the cost of carbon abated through this process simply by assuming the wind energy displaces an equivalent amount of gas and coal generation. No allowance is made for reserve requirements, or of the cost and loss of efficiency from retailoring the network to accommodate wind. On those favourable assumptions the report calculates the cost of saving each tonne of carbon at 54 Euros or several times the price of carbon on the European ETS at the time of the report. The cost of using PVs to reduce carbon comes to a staggering 716 Euros a tonne. The report concludes that it would be better for Germany to abandon subsidies for renewable energy and leave abatement up to the ETS market.

Despite Australia's decision to adopt a target of 20 per cent of electricity coming from renewables in 2020, very few studies have been done on the effects of adopting such a target and certainly nothing that can be considered comprehensive and independent. In essence, Australia has adopted a renewable energy target with no assurances about how much will be saved in emissions, if anything at all, and what it will cost to save those emissions. Nor is it anyone's job to work this out. The job of the grid operators is to make the mix work, while relying on the individual generators to supply the power. The distributors have to meet the legislative requirement which they will do. They don't have to work out whether any emissions have been saved, so they won't. The legislation allows for regular reviews and it is to be hoped that the government of the day insists that review includes calculations of costs versus emissions saved.

One potential bright spot for wind advocates in this sorry picture is the performance of sophisticated wind forecasting systems which are now being tested around the world. Perhaps these will improve the trade off for wind systems and give them a pretext for dismissing all this draining scepticism? Results of a system being tested in Australia handed to me, show that for five minute forecasts the system has an accuracy of more than 95 per cent at certain locations. That is the system will forecast the strength of the wind five minutes ahead and will be right more than 19 times out of 20. This sounds impressive until you ask yourself what forecasting accuracy you would achieve if you simply declared that the wind will be the same in five minutes as it is now. That simple "status quo" system of forecasting will work well until the wind changes. The figures for one hour ahead, four hours ahead and 24 hours ahead are also impressive at mostly more than 90 per cent, albeit with considerable variations between states, but just how effective this will be in reducing general uncertainty and reserve requirements is a question for the engineers. Those sorts of uncertainties still sit oddly with the need to maintain a continuous supply of electricity with steady voltage and frequency at all locations in the grid. The E.On reports and other reports I have seen mention that wind forecasting is improving but that it remains a difficult area.

Access to major hydro facilities would a help a lot, as the use of Norwegian dams has helped the Danes, but hydro-power is limited in Australia and excess power cannot exported as it is in Denmark.

This discussion of wind has not really touched on other, major problems with wind towers such as the ease with which they kill birds, the noise generated and their tendency to spoil the look of the landscape. The latest generation of wind generators, in particular, are enormous installations, difficult to ignore. News reports from the UK regularly say that wind towers in new projects will be taller than Big Ben, the clock on top of the Houses of Parliament in London. This problem may be partially solved by putting many of the new installations well offshore, but that makes the electricity more expensive, and increases the risks. Atlantic storms are fierce and the wind generators must be built to withstand the worst.

We also have also barely discussed other renewable resources such as photovoltaics, solar concentrators of different sorts including solar towers, and installations that harness wave power. There is plenty of coast on which to site wave-power generators and no shortage of sun in Australia, but every indication is that these projects will be yet another way to generate extremely expensive, although perhaps somewhat more reliable, electricity.

Solar concentrators generally consist of curved mirrors which concentrate the heat on a tube containing anything from water to oil. Spanish companies have built concentrator plants with mirrors focused on a central tower which uses molten salt to store heat, and I have seen media statements that those plants can be used as base load generators. Those reports typically say nothing about the cost of the electricity they generate. Far less elaborate versions of concentrator plants have been used very effectively to supplement hot water heating systems at hospitals, and may even be almost worth the cost of installing.

Two other forms of concentrators also involve towers. One approach, which counts as a proposal as no-one seems to have built one, involves a very tall tower surrounded by mirrors which concentrate light on the top of the tower. The extra heat creates an up-draft which can be used to power turbines. The concept is at least entertaining and there are plenty of places in Australia to site such structures, but the capital costs are substantial. Another involves using mirrors to concentrate light on a photovoltaic array, as the additional light makes the array more efficient. Two, separate proposals to build towers near the Victorian city of Mildura, one of each technology, have come to nothing.

Wave power comes to the fore every now and then and several suggestions have been made over the years to build a string of wave power stations along the northern coast of Western Australia. There is no technical reason preventing this. Wave power energy is more concentrated than that of wind and more reliable, albeit far from infallible, but economics and transmissions distances get in the way. Wave power is much more expensive than wind and that area is a very long way from major population centres.

One potential energy source often mentioned approvingly in Australia is that of geothermal and, at the time of writing, there are several companies trying to exploit this resource. This technology involves drilling down several kilometres to volcanically heated rocks, then circulating water down there to produce steam to drive turbines. This technology has the advantage of not being intermittent and so will have high value if and when it can be made to work. There are plenty of technical difficulties so it is best to come back to it when the technology is proven and ready to go. Another problem is that all the potential sites are in remote areas, which will require high capacity transmission lines stretching across parts of the outback.

What about smart networks with computer chips throughout the network that communicate with one another, and with the electricity suppliers so that loads can be shifted around more efficiently? Perhaps this network can be combined with a photovoltaic array on the roof of each house? Sure but again we run into the problem of cost and intermittency as clouds can drastically cut photovoltaic power. Grid operators will need a cloud forecasting system as well as one for forecasting wind, or lay in more reserve power. The overseas reports on photovoltaics are far from encouraging. The technology does not seem to be suitable for the mass production of electricity.

When renewable energy activists are tired of talking about the advantages of green power in avoiding carbon emissions, they will start talking about the jobs to be created and new industries to be founded if we wholeheartedly embrace the clean and green world of renewables. These claims are nonsense. Replacing 20 per cent, or 16 per cent, of the nation's power supply with a vastly more expensive means of producing electricity increases everyone's costs and that will mean job losses. Will there be more jobs in building and installing wind turbines, as opposed to building the equivalent capacity in fossil fuel stations? Whatever the result of the trade off, any extra jobs have to be paid for by electricity consumers, and the resulting increase in prices will cut jobs elsewhere. To claim that jobs will be created by drastically cutting efficiency in one sector, while saving little in the way of emissions, is like a physicist claiming to have invented a perpetual motion machine. It just isn't going to happen.

The Danish and German reports cited above go through a lot of economic analysis to reach the same conclusion and the Spanish economy, which has embraced all sorts of green solutions, has notoriously high unemployment. To be fair, however, that unemployment has far more to do with the bursting of that country's recent housing bubble. As for exploiting export markets, the problem is that activists in all the other Western countries are also talking about supplying export markets. Everyone will be exporting, it seems, and no-one will be importing. The Danes are better off at least in that respect, as they were among the first to be involved in wind energy and so now have a subsidised export industry. Australia does not need an interest in wind power to create heavily subsidised, inefficient export industries.

As I have discovered none of this impresses wind activists. Like their climate change cousins they are quite capable of dismissing vast amounts of evidence and reasoned argument on the slimmest of pretexts, although to date they have not had to work hard in dismissing the opposition. As noted, the enthusiasm for wind does not really affect the interests of the energy industry, and individuals in the power industry who may want to speak out suppose, correctly, that they will be attacked bitterly for stating the truth. The group that will be seriously affected, the consumers, have yet to see the bill. In the meantime a whole new interest group has been formed, the wind energy industry, which has no interest in a frank discussion of the costs and benefits of wind.

One indication of the power of this new wind group is a report entitled *Electricity from Renewable Resources – status prospects and impediments* produced by the US National Academy of Sciences in 2009. The US academy is, of course, a major and respected scientific body but the report somehow ignores or barely mentions the host of problems set out in this chapter. In its summary, the prepublication version says, "As a result of its study, the panel found that technologies for generation of electricity from renewable resources represent a significant opportunity – with attendant challenges – to provide low carbon dioxide emitting electricity generation from resources available domestically and to generate new economic opportunities for the United States."

This is another example (there are others) that many of the prestigious and supposedly independent scientific bodies around the world have gone "green" and can no longer be relied on for objective advice. To judge from some of the reaction online even the activists were taken aback at the report's tone.

As shown in this chapter wind energy is vastly more expensive than the general public realises and the savings in emissions likely to be negligible, but few reports have comprehensively examined the trade off in costs and emissions. With even supposedly independent organisations allowing themselves to be subverted to the extent of producing what observers may consider to be sales brochures, wind energy continues to rise and rise.

One coal industry executive commented to me that the adoption of wind energy bore similarities to the reasons for building the Great Wall of China. The Great Wall was built ostensibly to prevent northern tribes constantly raiding the border provinces but it is much too big just for that function. Instead, he says, the wall was really Imperial propaganda. It reminded the peasants that the Chinese government was massive, and not to be trifled with.

The analogy with wind energy is not quite right, as wind farms are being set up everywhere with at least nominal support from the mass of voters who have yet to see the bill. It is as if the Chinese peasants had welcomed the announcement of the wall without realising that they would have to build it. Wind farms may avoid some emissions at great cost, but like the achievements of the Great Wall, which also did not work very well, that does not matter. They are symbols and, in this unreal world of climate change lunacy, those symbols matter.

9
SEA, FLOODS AND ICE

One of my personal, laugh-out-loud moments concerning the endless talk of rising seas was when Sydney's *Daily Telegraph* named the Shire of Hornsby where I live as in danger from rising seas (*Daily Telegraph*, 15 May 2009). Most of the shire in Sydney's north is on a plateau well above sea level and my own house is on the side of a ridge, so I doubt that the local government planners are unduly worried. However, there is some, slight cause for concern in future decades if, as the article suggested, sea levels will rise 40 centimetres by 2050 as part of the shire fronts onto the Hawkesbury River. Although it is hardly a thickly settled area – most of it is bush – there are still settlements and various aquaculture businesses that may have to adjust if the forecasts have any basis in reality.

The dire fate of Hornsby is typical of the problems developed countries will face over the next few decades. Hornsby planners will have to adjust plans for foreshore development and that will take many meetings. In America, where there is a great deal more development closer to the shore line (as is my general impression), the forecasts are a little more serious. In low lying developing countries, Bangladesh is frequently mentioned, perhaps it may be more serious again. A sea level increase of one metre is supposed to be enough to flood half the country.

However, these forecasts mentioned above are a world away from the more extreme increases of six metres (18 feet) featured on the now infamous Al Gore 2006 video *An Inconvenient Truth*, already mentioned once or twice in these pages. They also sit oddly with frequent warnings that the Greenland ice shelf is about to disappear; the Arctic is vanishing; the Antarctic is breaking up; glaciers everywhere are vanishing; the weather has been very odd lately. It's all the fault of global warming. We are all doomed!

So should we pay much attention to these milder forecasts and what happened to the watery Armageddon we were promised in 2006? A quick, common sense review of the area indicates that we should ignore all these forecasts, at least for the next decade or so. Some increase in sea levels is inevitable, but the ordinary, long-lived beach goer may never notice them. The real problem is and will remain coastal erosion in places where people have been putting buildings for the last century. Sea level increases may accelerate that erosion in a few decades to come but that has yet to be proven. I believe some scientists and all the activists would dearly like to beef up these forecasts and return to warnings of multi-metre inundations, preferably happening tomorrow rather than next century, but they are stopped by reality. There is now general agreement that ice sheets do not break up that easily, or that quickly, should temperatures actually increase. The process takes time.

First some history. The 2001 IPCC report forecast a range of increases in sea level of up to perhaps a metre over a century, with the bulk of the increase expected to result from the topmost level of the ocean becoming warmer – a body of water expands as it warms – plus additional run off from glaciers and the enormous Greenland and Antarctic ice sheets. (The Arctic does not count, despite changes in the summer ice there in recent years, as it is all basically a giant ice cube floating in water. Melting it will have no effect on sea levels.) The 2007 report was milder, at least as far as sea levels are concerned, forecasting increases of between 0.18 and 0.59 metres by 2100. That 0.59 metres estimate, incidentally, is for a temperature increase of 6.4 degrees, which is well above the 4.5 degree maximum given for the surface temperatures in the same report. Let us pass over that puzzle to the resulting fuss raised by scientists who considered the forecasts far too conservative. They lobbied hard for a revision, pointing to alleged increases in the rate of melting of glaciers. Eventually the forecast was pushed up to a maximum expected sea level rise of 0.76 metres for the century. After extensive searching I am not clear as to exactly who is forecasting this, or whether the IPCC explicitly changed its forecast, but that is the figure that seem to have been handed to the Hornsby council planners. Half 0.76 is just under 0.4 metres in 40 years which, incidentally, is about one centimetre a year or three times the present rate of increase.

A paper in the journal *Nature Geoscience* managed to push the maximum increase out a little further but, unlike much of the screaming about collapsing glaciers and ice sheets, seems to be a useful contribution to the debate. The paper, actually a letter to the journal ('Constraints on future sea-level rise from past sea-level change', 26 July 2009) reports on efforts by Mark Siddall at the Department of Earth Sciences at the University of Bristol and others, to reconstruct sea-level changes over the past 22,000 years from fossil data, and match it to known temperature changes. As is widely accepted during the coldest part of the last ice age about 20,000 years ago, with a lot of water locked up in vast ice sheets, sea levels were about 120 metres lower than today. One of these vanished chunks of ice was the Laurentide ice sheet which covered a good part of North America. When a warming climate ended that ice age sea levels rose sharply from about 14,000 to 7,000 years ago to somewhere near present levels and have been increasing slightly over time ever since. Other work, which we will get to in a moment, suggests that there have been pauses and accelerations in sea level changes.

After matching those known changes to known changes in temperatures and plugging in the IPCC forecasts, Siddall and colleagues came up with a likely increase of 7-82 centimetres in sea levels by 2100 which they note is higher than the increase of 18-76 cm given by the IPCC models, but not by much. As it's a simple and fairly robust model based solidly on past trends rather than computer projections, it is reasonably persuasive. Waving aside the problem of whether the IPCC temperature forecasts are correct for the moment, accepting the Bristol work saves us a lot of trudging through papers and presentations by scientists determined to "prove" that the run off from this glacier or that ice sheet is increasing. As we noted in a previous chapter, the various blocks of ice dotted around the earth act as a measure of temperature, with major local variations. Temperatures were increasing between the mid-70s and the turn of the century and that increase is probably still working its way through the glaciers and ice sheets. That means ice sheets still are shrinking, but there is no way to work out whether this shrinkage is in any way unusual or due to industrial activity.

An increase of well under a metre in a century is not going to galvanise policy makers, grab headlines or bring in dollars for research grants. So scientists are always looking hopefully for ways to bump up the likely increase. One group found some hope of additional sea level increases in its study of the break up of the previously mentioned Laurentide ice sheet at the end of the last ice age. The ice sheet took some time to melt away with researchers pointing to two major melting phases, one 9,000 years ago and the other 7,600 years ago. In a paper ('Rapid early Holocene deglaciation of the Laurentide ice sheet', *Nature Geoscience, 31 August 2008*), Anders Carlson of the University of Wisconsin-Madison and colleagues push for a possible increase of 1.2 metres per century, saying that the break up of the Laurentide ice sheet happened that fast and present conditions are analogous.

Perhaps; although there is still the problem that those projections depend on the IPCC forecasts being somewhere close to reality. As we have seen that is asking a lot. But even if the IPCC forecasts have some validity is there any problem in waiting a decade, perhaps two, to see just what will happen given the small increases involved?

Global warmers would react to such a suggestion with horror, and with claims that rise in sea levels is accelerating. Such claims were made at a climate change congress in Copenhagen in Denmark in March 2009 and repeated in a document entitled the *Copenhagen Diagnosis* produced in the following November by a group of concerned scientists, just before the international meeting in the same city on reducing emissions. Although the document made a range of claims about how everything was getting worse faster than expected it generated surprisingly little publicity, which may be a sign that the public is becoming inured to warnings of apocalypse.

In that document, the concerned scientists say that sea levels have changed by about 20cms over 130 years. Sidall and colleagues quoted above give an estimate of 20cms for the entire 20th century. But the concerned scientists also point to satellite data which gives an increase of 3.1 mm a year, for 1993 to 2008. If that increase continues for a century sea levels will increase by 0.3 metres or about a foot in the old Imperial measures. This they compare with projections from the 2001 IPCC report which forecast sea level increases of 1.9 mm a year (0.19 metres over 100 years, or about in the ball park for the increase over the previous century).

This is curious as data from the Topex/Poseidon satellite on which they rely, a joint venture of the NASA's Jet Propulsion Laboratory and the French space agency CNES, started operating in the early 1990s as the data would suggest. Its task has been taken over by the more recent Jason satellite. The satellites use a combination of laser tracking and radar altimeter to continuously check sea heights beneath them. The scientists who compiled the IPCC 2001 report should therefore have been aware of the project and its results, which do not seem to have changed substantially in that time. The data, which can be accessed in graphical form on a site run by the University of Colorado, does not show much variation from the trend line of 3.1 mm a year. Sometimes the data is a little above the line, sometimes below. In the last few years the data has been below the line.

The report does not refer to any integrated line of satellite or tide gauge survey results showing an acceleration in the rate of increase in sea levels, but there is recent research work which reconciles the 2mm a year rise inferred from tidal gauge observations with the higher satellite measurements. Unfortunately, the reconciliation will not be to the liking of the global warmers.

Accurate records of tidal gauge have been taken in some locations for centuries, but they are not easy to interpret. They can be complicated by all sorts of local factors such as tides, storms, harbours silting up or beaches washing away, not to mention the ground itself sometimes subsiding, or uplifting very slightly due to volcanic activity. Ground that was underneath glaciers in the last ice age is still rebounding after having the ice on top of it melting away. Those changes are not very large, but then neither is the increase in sea levels. Nonetheless a group of researchers from the Proudman Oceanographic Laboratory in Liverpool in the UK and the Arctic Centre at the University of Lapland in Finland, analysed whatever gauge data they could find to show that the rise of about 20 cm during the 20th century was far from uniform. In a paper 'Recent global sea level acceleration started over 200 years ago?' by Jevrejeva and others (*Geophysical Research Letters*, 30 April 2008) they present a reconstruction indicating that the rate of global sea level rise increased and then decreased in several distinct waves since before the beginning of the 19th century, or about the end of the Little Ice Age. They say there is a background acceleration which started 200 years ago (well before emissions kicked in), with another cycle imposed on top of that, meaning there are periods of several decades when the rate of increase is above and below normal. The average for the 20th century still works out to 2mm a year or so, but we are now at the top of one of the cycles. The group's calculated rate of increase of 3.4 mm a year is a good agreement with the satellite data of 3.1 mm a year. The rate of increase was at least that high in the 1950s.

This work is still fairly recent and has to be kicked around scientific circles some more, but it does not seem to have been challenged and has been taken as supporting the IPCC. In the paper Jevrejeva and colleagues say that simple extrapolation of the results gives a sea rise of 34cms during the 21st century. However, the paper says that even if temperatures rise more slowly than the IPCC suggests, "oceanic thermal inertia" and rising Greenland melt rates may push sea levels along faster than the panel's projections.

Arguing that point is well beyond where we want to go in this lay person's tour of the science, but we can see that a good part of the acceleration in sea level increases cited hopefully by the scientists at

Copenhagen may well be due to natural variations, and not a great deal is left for the supposed vast amounts of additional water pouring off glaciers and ice sheets. It is interesting that the scientists at Copenhagen did not draw attention to the Jevrejeva paper, although it does reconcile the satellite and tidal gauge data nicely, or to the Sidall material. All of the above is still subject to argument, of course, but none of it really adds up to more than 0.8 metres or so over a century, and may add up to a lot less.

While on the subject of sea height increases we should note the plight of the inhabitants of the Carteret Islands, about 90 kilometres north east of Bougainville in the Southern Pacific who, reports indicate, are being washed away. Growing food on the islands, which are supposed to have a maximum elevation of 1.5 metres, is becoming increasingly difficult. The small increase for all of the last century should not have caused such problems but local increases or declines in sea level can be substantial. One among a number of suggestions for the source of the problems faced by these islanders is that local movements in the earth's crust are causing the islands to subside. The last people to see this issue in a balanced way, however, are activists who are billing the 1,000 or so Carteret islanders as the first wave of climate refugees.

After examining the figures for sea level increases, and those for temperatures and carbon dioxide concentrations in previous chapters, and been unable to find cause for concern, we will not spend much time on allegations that the world's ice is melting away much faster than expected. As noted above, there may well be some additional melting due to present temperatures being historically high, but a supposed acceleration in that melting does not seem to be having much effect on sea levels. However, our ever hopeful activists have been encouraged by changes in the Arctic.

As all the world knows the amount of sea ice around the Arctic has dropped in the last three decades, and this has been interpreted as a sign that the earth is heating up and we are approaching a tipping point. Figures from America's National Snow and Ice Data Centre shows that Arctic sea ice has shrunk by 5.9 per cent a decade for the past 30 years, with sea ice coverage dipping sharply below that trend line in 2007, to recover in subsequent years. The regular cycle of ice forming in the sea around the Arctic during winter and then melting during summer also occurs around the Antarctic, but that cycle does not show the same change. We know this because the sea ice cycle has been tracked by satellite observation with reasonable accuracy at both Poles since the 1970s. Whatever the reason for the undoubted change in the Arctic's annual freeze and melt cycle, vessels not designed as ice breakers can get through the fabled Northwest Passage. During the Little Ice Age European navigators tried to force the Northwest Passage from the Atlantic around the top of Canada into the Bering Sea and North Pacific, but found the way blocked by ice.

Activists and some scientists are in no doubt as to the cause, and have seized on this undoubted change in the Arctic sea ice cycle as proof that the earth is warming dangerously. In the process some of the more enthusiastic activists have confused changes in sea ice cycle with a melting of the Arctic ice cap, then puffed the observations up into claims that whole Arctic will disappear in a matter of years. The problem with claims that the Arctic itself is getting smaller is that there is little, reliable data for either Pole on which to base that judgement, and certainly not reliable data sets over decades or centuries. Platoons of scientists are now wandering over both Poles and poring over satellite data looking for shrinkage, but then they keep on rushing into print with different stories. Matters have been further confused by various lurid reports of this or that piece of Antarctica melting or collapsing over the years, notably the Wilkins Ice Shelf at the base of the Antarctic Peninsula, a small tail jutting out from the mass of the continent due South from the tip of South America. The problem with these reports, as sceptics point out, is that the Wilkins Shelf has been reported as collapsing or near collapse "for the first time in history" on a number of occasions over the years. After checking newspaper and internet archives I believe the Wilkins Ice Shelf was first reported as collapsing in the 1990s. One sceptic site recently complained that the same photographs are reused each time the story is reported.

In any case, as we have noted several times, because temperatures were undoubtedly rising until at least the turn of the century and are still historically high it would be far from surprising to find some ongoing melting. Also, as we are already directly measuring changes in the earth's temperature there is little point in concerning ourselves with an indirect measure such as changes in ice volumes and coverage, although it is always nice to know these things.

But then what about the sea ice in the Arctic, and why isn't the Antarctic affected? Scientists have been casting around for explanations for the difference. Massive ice shelves jutting out from the Antarctica coast and extending under the sea ice have been discovered recently. Perhaps, New Zealand scientists suggest, those shelves are protecting the Antarctic from the warming trend evident in the Arctic. Another suggestion is that the hole in the ozone layer is channelling a mass of cold air over Antarctica keeping it cool.

Another possibility is that the changes in the Arctic sea ice are only slightly related to temperatures. When *The Guardian* newspaper in the UK reported in 2007 that the Northwest Passage was nearly ice free for the first time since records began, it meant since the US National Snow and Ice Data Centre started routine satellite monitoring in 1972. The route has been open before. The main feature of one small museum in the Canadian city of Vancouver is the Saint Roch, a small wooden vessel with some metal on its bows and a 150 horsepower engine. One of the major claims to fame of this vessel is that in the 1940s it traversed the Northwest Passage in a single season, and is supposed to have done so three times.

One possible explanation for the apparent variations in ice in the Northwest Passage has been put forward by Syun-Ichi Akasofu, of the University of Alaska Fairbanks, in Fairbanks, Alaska. A founder of the International Arctic Research Centre at the University and a strident critic of the IPCC, he says that the sudden decline in sea ice is due to a cyclic change in ocean currents. In what he describes as a note entitled 'The Recovery from the Little Ice Age' put on line Akasofu says that the change is the result of an influx of warmer water from the North Atlantic due to the North Atlantic Oscillation (we encountered the NAO in an earlier chapter). The thinner ice that resulted from that inflow was more easily broken up by storms. He also points to research that the shift is a cyclic pattern and that the changes in the sea ice are not uniform, with a larger change occurring along the Siberian coast where more of the warmer NAO water has flowed.

This book is not about trying to adjudicate between those various claims, particular as neither the Akasofu nor the New Zealand material seems to have been formally published, although I am more impressed with Akasofu's reasoning. The point is that we do not know a lot about what is happening at either Pole and, in any case, we have other ways of measuring global temperatures so we should return to those. As for the supposed rapid melting of the glaciers, just before this book was written the IPCC was caught in one of its most embarrassing errors to date which concerns glaciers.

In its 2007 report the IPCC says that glaciers in the Himalayas are receding faster than in any other part of the world and at present rates of shrinkage are likely to disappear by the year 2035 or perhaps sooner. As the panel has now admitted, the estimate was based on a World Wildlife Fund report, and the WWF report was based on a report in *New Scientist* in 1999. That story, in turn, was a report of a little known Indian scientist speculating about when the glacier might melt. (*The Sunday Times*, 17 January 2010.) The same *Sunday Times* article and another in the *Daily Mail* on the same date quoted glaciologists as saying that they were astonished that the claim had taken so long to be exposed. The Himalayan glaciers are so big that they would take centuries to melt, not decades.

The IPCC has since admitted that there may have been an oversight, but it is curious that the supposedly carefully checked report based a major assertion on a statement in a green organisation document. Although the error is not significant in itself, the admission does not encourage much confidence in the rest of the document. As this incident is still warm at the time of writing, so to speak, we will move on to the last stop in our investigation, the mechanism involved in icebergs breaking away from the enormous ice sheets of Antarctica and Greenland.

When scientists forecast the collapse or perhaps dramatic shrinkage of these ice sheets they are not suggesting that they will melt on the spot. Not even the most dramatic IPCC forecasts push temperatures around those ice sheets to above zero. Instead they are suggesting the process of calving will speed up. As is well known, icebergs break away or calve from the face of the ice sheets then float off into warmer waters where they melt. Their place is taken by more ice. Scientists refer to an ice sheet's mass balance, that is the amount of snow and rain falling on it, as opposed to the icebergs floating away from it. If it's losing more ice mass than it's gaining in precipitation then the ice sheet is contributing to sea levels; if it's gaining mass then sea levels should be falling. The process of calving, like most processes in nature, is more complicated than simply a chunk of ice snapping off. One of the variables that affect the process is the amount of water that gets to the base of the ice sheet, assuming it rests on land, which lubricates the slow downhill creep of the sheet. In other words the water makes it slip downhill faster, so increasing the rate of calving. Thus comparatively minor changes in temperature may cause big changes in ice sheets.

There is some argument over whether the calving process works that way, but the Greenland and Antarctica ice sheets are not going anywhere just yet. In an article in the Australian Institute of Geoscientists News 'Why the Greenland and Antarctic Ice Sheets are not Collapsing', (*Quarterly Newsletter*, No 97, August 2009) Cliff Ollier of the school of earth and geographical sciences at the University of Western Australia, and Colin Pain, point out with some force that much of the ice in Greenland and Antarctica occupies deep basins. That means they are not about to slide down a slope. They have some harsh words to say about ice sliding down a slope model, but also point out that both sheets have hundreds of thousands of years worth of accumulated ice with no melting or flow, and it's all still far below freezing point.

All this means we are still left with the previously mentioned undramatic sea level increases, which depend on actual temperature increases being close to the top end of the IPCC projections. As we have seen those forecasts may have nothing to do with reality and actual sea level increases are running at one-third of the amount required for even the modest panel forecasts. Scientists can and do add storm surges on to the projections to get a maximum of just over one metre, but it's still not really enough for a Hollywood blockbuster film.

Raising concern let alone panic over such increases is hard, and it is difficult to see what harm there can be in waiting a decade or so to see whether there is any acceleration in sea level increases, particularly as there is now a satellite tracking system delivering reasonably accurate information on the issue. An additional advantage of the satellite system is that we can look at the rate of increase for ourselves, rather than have the data filtered through an IPCC report.

Planning authorities may not want to wait for a decade or so before setting building codes or regulations concerning building near the foreshore, and if so the New South Wales state government forecast cited at the beginning of the chapter is as good as any other. Other Australian states, Queensland, Victoria and South Australia, have opted for a slightly smaller forecasts, of 30 cm by 2050, and there is not much wrong with that figure either, unless they plan to permit buildings expected to last for more than, say, 60 years, directly on the foreshore. The situation can be reassessed in a decade. The real problem for foreshore planning remains beach and coastal erosion, which is likely to act far faster and be more costly than any global sea level rise, if it occurs. The real lunacy of this chapter is that we may spend a lot of time obsessing over an almost non-existent threat and forget the real danger.

10
ACID AND ADAPTION

In early 2004 the major scholarly journal *Nature* published a paper which used the mid-range IPCC temperature projections to forecast that 15 to 37 per cent of species will be "committed to extinction" by 2050. The paper had an astonishing 19 authors all from prestigious institutions, led by Chris Thomas at the University of Leeds' centre for Biodiversity and Conservation in the UK, and was accepted for publication a month after being received in September 2003, ('Extinction risk from climate change', Chris Thomas and 18 others, *Nature*, 8 January 2004.)

These "terrifying" conclusions were constructed using the best computer models available, with the co-authors mainly supplying data on climate and species distribution for the models over 20 per cent of the earth's surface. Much of the subsequent media comment focused on how the paper had ignored the effect that changes in land use by humans and rise in carbon dioxide levels would also contribute to species extinction. With so many scientists and so much expertise behind it surely the paper must be right, and how dare a mere layman question it? Well pardon my lack of qualifications but the paper fails the common sense test. The reviewers and editors of *Nature* should have at least hesitated and requested a few caveats be added to the paper before publication.

As we have seen, for temperatures to reach even the mid-range IPCC projections by 2050 they will have to get moving. So this research is relying on one set of computer model projections, which may be quite wrong, to make another set of computer model projections. But even if we accept the IPCC estimates, would a degree and a half increase in temperatures in fifty years have the same effect as a major meteorite impact or the end of an ice age (see below)? Bear in mind that, also as we have seen, the Medieval Warming Period is known to have been at least at least a degree warmer in Europe without any medieval chronicler noting a sudden change in the number of animal species. As is well known, there were extinctions in medieval times but through the clearing of the great forests which used to cover much of Europe, rather than through any change in temperature.

A much better approach, and one that avoids the use of computer models, would have been simply to count up the number of extinctions that are known to have occurred between the mid-70s and the turn of the century when there was a marked surge in temperatures, and reason from that. The problem with that approach is that despite all the cries about species extinction and all the dreadful things humans are supposed to be doing to the world, there are few countable extinctions. Various species have gone extinct in recent years, such as the golden toad in Monteverde in Costa Rica, but in nothing like the numbers required to justify these warnings. In all the public statements about species extinctions very few of the concerned commentators mention species that have gone extinct, as opposed to those on a danger list.

As it turns out, these and other dire warnings about how birds and animals are about to start dropping dead any minute failed to take into account natural adaptability, a point made by Kathy Willis, of the long-term ecology laboratory at Oxford University in a paper in *Science* ('Biodiversity and Climate Change', Kathy J. Willis and Shonil A. Bhagwat, *Science*, 6 November 2009).

Willis, who declares emphatically that she is not a climate change denier, says the computer models used to forecast species extinction rates have very "coarse spatial scales" (that is each data point may be 16 kilometres by 16 kilometres) and do not take into account topography or habitat variations within each of those areas. If the models are run using a finer scale, 25 metres by 25 metres, the results are different. In addition, the models do not take into account the full "acclimation capacity" of plants and animals. "Several recent studies indicate that taking these factors into consideration can seriously alter the model predictions".

An article in *The Times* which quotes the Willis paper also quotes Keith Bennett, a professor and head of geography at Queens University Belfast (*Times Online*, 6 November 2009), comparing forecast extinction rates with actual rates. To be on track to meet the forecast figure of one-third of bird species to become extinct by 2050 Bennett calculates that 36 species of birds would have to become extinct each year up to 2050. In reality, between 2004 and 2008, three species of birds became extinct. In the article he says that bird species are far more versatile than some prediction models suggest. If it gets warmer, instead of dying out they move on.

All this also supposes that scientists really understand the factors that lead to extinctions. In his book *The Long Summer How climate Changed Civilisation* (Basic Books, 2004), former professor of archaeology Brian Fagan, says that between 14,000 and 9,500 BC as temperatures warmed and the steppe/tundra of Europe gave way to tree cover, a wave of extinctions affected animals in the area. The mammoth, woolly rhinoceros, giant deer and many smaller animals became extinct. You would think that the increasing temperatures had something to do with this, but apparently it's not that simple. Fagan says, "A variety of complex and little-understood environmental stresses led to the extinction of the more specialised and less adaptable Ice Age species."

The Australian versions of ice age mega fauna included a possum-like creature about the size of a rhinoceros, considered the largest marsupial ever, and a giant goanna-like meat-eater. The exact reasons these creature became extinct is unclear but it is thought a major factor were the drought conditions in Australia at the depths of the last ice age about 18,000 years ago.

All those extinctions occurred during very substantial changes in temperatures and conditions altering vast areas of the earth out of all recognition. Besides those changes, an increase in temperatures of less than a degree since about 1860 is barely worth mentioning, let alone the smaller changes of more recent decades. A change of a degree or two in perhaps just 40 years may be more serious, but estimates of the number of extinctions remain speculation based on unproven computer models. In fact, human activity is affecting birds, animals and plants more than they should but not through climate change. A point made by Willis cited above is that human changes in land use have had a far greater effect on species, and policies for managing our ecosystems should take that into account. Forget climate change; concentrate on land use changes.

This problem of concern over climate change, real or imagined, diverting attention from far more immediate and serious human-induced effects, can be seen in the debate on the possible fate of oceans and coral reefs. Coral reefs are masses of limestone made from the skeletons of countless millions of tiny marine animals and plants. Colonies of tiny, living coral polyps grow on a reef's surface. These animals extract dissolved calcium compounds from the water and, with the help of single celled plants (called zooxanthellae) living inside them, lay the material down as hard limestone around the lower half of their bodies. This is a continuous process but by definition the polyps can never be far from sea level.

The Great Barrier Reef and other reefs around the South Seas may be in danger from human activity but the danger from local mining, farming blast fishing (fishing with explosives which still sometimes occurs in the South Seas) or simply pollution being piped out to sea, is far greater than climate change. We know this because the reefs have already lived through changes in climate much faster and sharper than anything seen in the last century, or anything likely to come. A fact sheet for tourists produced by the Great Barrier Reef Marine Park Authority (which also supplied the brief description of reef building above), says that corals have existed in the area for 25 million years, and a complete reef structure such as those seen today existed in the area about 600,000 years ago. The present reef started growing on top of the old reef platform about 9,000 years ago, when sea levels stared rising after the end of the last ice age.

As you will recall from earlier chapters, 600,000 years covers six ice ages and interglacials, with perhaps at least two of those intergalacials featuring warmer temperatures than present, with all the resulting vast changes in temperatures and conditions in-between. As you will also recall some of those changes are known to have been faster than anything experienced recently. At one point in the great rise in sea levels between 14,000 and 6,000 years ago those levels are known to have risen at a rate of 40 millimetres a year, or an order of magnitude higher than the present rate. Carbon dioxide concentrations also increased sharply in the same period, albeit lagging behind temperatures. Given all that it seems unlikely that a change of less than a degree in 150 years, and recent increases in CO_2, will be too much for the reef. Activists respond that higher temperatures are expected soon, but in the unlikely event those forecasts are realised the reefs will still have survived a lot worse.

A book *CO_2 Global Warming and Coral Reefs: Prospects for the Future* by Dr Craig D. Idso has a lot to say about how reefs and marine life can adapt. The book is produced by the Centre for the Study of Carbon Dioxide and Biological Change and the Science and Public Policy Institute, which means that activists will immediately want to dismiss it out of hand as the SPPI are the enemy and activists see matters of science in political terms. As we are more broad-minded we will note that Idso's book lists a considerable number of refereed studies on coral reef adaption. He also says that if sea levels do start increasing substantially then the reefs may benefit from this. As we have seen sea levels increased by 20 centimetres over the last century or so, which is very little in geological terms, and the reefs have responded to the relatively stable sea levels by building out. Idso says that a faster increase in sea levels, should it ever occur, will give the reefs a chance to build up.

A little more worrying is that Idso also says that there has been a build up of coral reef bleaching during the 1980s and 1990s. This has been blamed on global warming (what else?) although as we have seen this seems unlikely. Periods where ocean temperatures have been degrees higher and lower due to atmospheric cycles such as La Niña and El Niño, are known to cause bleaching from which the reef recovers, but that cannot be causing the more widespread episodes. Scientific investigations into the likely reason for this bleaching are far from complete but one of the leading suspects is an unusually prolonged drought in the Sahel region of Africa, the bit of North West Africa below the Sahara. That drought, due to a change in the ocean and atmospheric circulation systems, has resulted in a gradual increase in the dust content of the atmosphere. That dust carries bacteria, viruses and fungi that can spread coral diseases and there is evidence linking dust load in the atmosphere with coral disease. That is one theory, take it or leave it.

As this book was being written senior Australian marine researchers studying the Great Barrier Reef, including an episode of severe bleaching around the Keppel group of islands, were reported as saying that the reef has recovered from the bleaching episode and shrugged off a variety of human threats, including various poisons in run off from agricultural land. These scientists are Peter Ridd, a physicist at James Cook University in Townsville, and Ray Berkelmans at the Australian Institute of Marine Science, (*The Australian*, 18 December 2009).

Kill off one alarmist story, and another one is ready to take its place. The next disaster story in line is generally dubbed "acid oceans" and, despite obvious problems almost as large as those with the heat-killing coral theory, has proved reasonably popular among scientists seeking funding. As you will recall the concentration of CO_2 in the atmosphere has increased from about 330 parts per million in the mid-70s to just short of 390ppms in 2009. This change in concentration, from 0.033 per cent of the atmosphere to 0.039 per cent, means that there is more carbon dioxide to dissolve in sea water. CO_2 combines with trace chemicals in the water, bicarbonate and carbonate, to become the weak carbonic acid. One result is to turn the top most layer of the ocean slightly more acidic. Further changes in CO_2 concentrations should cause additional slight increases in ocean acidity. A discussion of acidity would fill a whole book but all you really need to know here is that chemicals are assigned a pH reading from 14 which is very alkaline or basic, depending on the terminology you use, through to 0 in theory although some scales show pH going out to -5. Bleach has a pH reading of 13, soap rates a 10, battery acid has a pH of 0 and hydrochloric acid is rated as -1.

Ocean water has an average pH of just under 8.1 but a graph in *Position Analysis: CO_2 Emissions and Climate Change: Ocean Impact and Adaption Issues* produced by the Antarctic Climate & Ecosystems Cooperative Research Centre in Hobart, Tasmania, in 2008 shows that pH readings in sea water can range from a bit below 7.8 up to a little under 8.5 with the bulk clustering around the 8 to 8.2 range. The pre-industrial level, that is the level in the mid-18th century, averaged just under 8.2. The forecast average for 2100, according to the centre, is a touch above 7.7. As pure water has a pH of 7 we are not talking about fish swimming around in acid, as some of the publicity on this issue suggests. Swimmers would never notice any of these changes. In fact, the actual acidity of the oceans is mostly not the problem, it's just a handy tag for referring to the problem.

Scientists say that these chemical changes will lead to other chemical changes which will affect the formation of a key chemical, calcium carbonate. Because there is less calcium carbonate floating around it will not be available to very small organisms to form shells, particularly microalgae and forms of plankton and the like. There is a whole zoo of these tiny creatures. Coral reefs will also have less calcium. A special issue of *Current*, the Journal of Marine Education, entitled *Ocean Acidification From Ecological Impacts to Policy Opportunities*, produced in 2009, says that the changes will affect the Coccolithus, a single celled species of plankton which has its own equally tiny shell. This creature is much smaller than a grain of sand but is still a key part of the oceanic food chain. In an experiment cited in the Antarctic CRC paper scientists grew one batch of Coccolithophorids (lots of Coccolithus) under laboratory conditions at pre-industrial CO_2 levels and another batch at three times pre-industrial levels. Those grown at the much higher CO_2 levels came out with malformed shells. From these sorts of experiments, scientists suppose that there will be major effects on the food chain.

There may be other effects. Baby clown fish, or at least larvae somewhat younger than Nemo in the popular film *Finding Nemo*, use smell as a clue to find a reef in which to set up shop as an adult. In an experiment at James Cook University in Townsville, Queensland, scientists raised Clownfish larvae in acidity levels expected at the turn of the century and found that the baby Clown fish's sense of smell was greatly affected. From that they have theorised that changes in CO_2 levels might disrupt population sustainability, ('Ocean acidification impairs olfactory discrimination and homing ability of a marine fish', Philip I. Munday and others, *Proceedings of the National Academy of Sciences of the USA*, 2009.)

As quickly pointed out and freely admitted in all the literature, there is still the issue of fish and smaller creatures simply adapting to comparatively gradual changes in CO_2 levels. As we have seen from the example of the birds cited above, creatures of all sorts have considerable adaptive powers, and they will not just be taken from one environment and thrust into another. Scientists then argue that the increase in CO_2 is faster than the capacity of sea creatures to adapt (*The Guardian*, 10 March 2009), although there is no real evidence of this nor could there be. The fossil record is not that exact. In any case, the experiments cited above use an astonishing CO_2 level of 1,000 ppm forecast by the IPCC. As we have seen carbon dioxide levels will not reach that figure for a long time yet.

Besides the problem of creatures adapting, there is the problem of trying to work out what would happen in a natural system that varies enormously. The acidity of the future ocean will be far from constant, varying a good deal around the average, depending on local conditions and currents and the like. To find out just how ocean acidity will affect clown fish or the plankton mentioned above, we have to rely on yet more computer models simulating the spread of acidity and its interaction with the habitat of various sea creatures. We have already discussed the problems of applying computer models to marine environments. The scientists involved also freely admit there are vast gaps in their knowledge about the factors involved. In other words we are dealing with projections based on projections, which are being used to make further projections, all of which involve major uncertainties.

One further possible cause for concern is the work of a team from the Antarctic CRC cited above, which collected the shells of one foraminifera species as they drifted towards the sea floor and then compared them with specimens that had sunk several centuries earlier. The recently deceased creatures had shells with 30 to 35 per cent less mass indicating that there has been change since the industrial revolution, ('Reduced calcification in modern Southern Ocean planktonic foraminifera', Andrew Moy, William Howard and others, *Nature Geoscience Online*, 8 March 2009). The paper also says that the team was able to find a link between higher atmospheric CO_2 and low shell weights in a 50,000 year long record obtained from a Southern Ocean marine sediment core. However, the paper also notes that it is unclear whether reduced calcification will affect the survival of this and other species. Those who care to look can find a lot more of that sort of evidence, an indication here, a projection there, but no immediate, tangible effect on fish stocks.

All of that adds up to a cause for concern, worthy of perhaps more research and paying a platoon or so of scientists to keep a close watch on the issue, but it is not the biggest threat to the world's fish resources. It is not even close. Over fishing has had much greater and far more immediate effect than anything forecast from industrial emissions. As we saw in an earlier chapter the enormous Grand Banks fishery, one of the richest in the world in its time, collapsed in the early 1990s due to bad management. Plenty of others have collapsed or are in danger of doing so, in what is developing into a serious problem for the fish industry world-wide. In 2006 Boris Worm, an assistant professor of the biology department at Dalhousie University in Halifax, Canada, graphed the rate at which fisheries were collapsing and forecast that they would be all finished by about 2050, ('Impacts of Biodiversity Loss on Ocean Ecosystems Services', Boris Worm and others, *Science*, 3 November 2006).

That rather grim forecast has since been revised, and scientists say it is avoidable. One relatively easy solution is to adopt a different approach to commercial fishing, by allocating fishermen and women transferable property rights in their fishery. The quotas are set by a central government body but the rights to a quota can be bought and sold. This general approach is used widely in Australia and seems to encourage individuals to think long-term about the resource in question, although my impression is that the government still has to be firm in setting quotas, ('Can Catch Shares Prevent Fisheries Collapse', Christopher Costello and others, *Science*, 19 September 2008.) Other papers have suggested that the matters are improving but, of course, governments could always do more about the issue and do it sooner.

The problem would be given an enormous boost if senior scientists also developed some sort of perspective on the issues, or read any material on the fishing industry. But instead of talking about the need to curb over-fishing as well as urge that the effect of increasing CO_2 levels on oceans be watched carefully, a statement by the Interacademy Panel, a peak body for the national science academies based in Trieste, Italy, thumps the alarm button as hard as it can.

The statement says that the rapid increase in CO_2 emissions since the industrial revolution has increased acidity in the oceans with "potentially profound consequences for marine plants and animals"'. Further, "at current emission rates models suggest that all coral reefs and polar ecosystems will be severely affected by 2050 or potentially even earlier". This statement was endorsed by a string of science academies around the world including the Australian and US academies of science, seemingly without question. Martin Rees, president of the Royal Society in Britain, went further in newspaper interviews. In an interview with the *Sydney Morning Herald*, (2 June 2009) Rees, an astronomer, says that unless global carbon emissions are cut by at least 50 per cent of 1990 levels by 2050 there may be an "underwater catastrophe" and loss of marine life.

A lot more statements about looming marine catastrophe can be found by those who go looking for them. A statement dubbed the Monaco Declaration, as it was made at the Second International Symposium on the Ocean in a High-CO_2 World held in Monaco in

October 2008, says that "ocean acidification is accelerating and severe damages are imminent". Signed by 155 scientists from 26 nations and based on "irrefutable scientific findings" the statement says that "along with increasing emissions, the increase in atmospheric CO_2 is accelerating. By mid-century the average atmospheric CO_2 concentration could easily reach double the pre-industrial concentration. At that 560 ppm (parts per million) level it is expected that coral calcification rates would decline by one -third". Reefs would erode away and so on.

Again, as we have seen, these scientists talking about "irrefutable" evidence are relying on IPCC CO_2 projections which are questionable to say the least. For CO_2 levels to get from a shade under 390 ppm to 560 ppm in just 40 years – a 40 per cent increase – is a very serious stretch. The many talented scientists who signed the declaration talking about irrefutable scientific findings, should at least have looked at trends in CO_2 concentrations for themselves.

In defence of the scientific academies and the Monaco scientists, carefully worded, cautious statements simply get lost in the ether. Unless you are talking about a catastrophe no-one will listen. There are no penalties for making such statements, and some rewards. Activists glory in dire warnings, often embellishing them, and the media dutifully report them. Organisations hand out awards to the scientists who make them and, last but not least, politicians may be harangued by activists into giving more funding to scientists with good scare stories. Look at the career of Paul Ehrlich, the professor of population studies at Stanford University in the US, whose 1960s forecasts of mass starvation in the 1970s and 1980s due to population outrunning the food supply proved to be quite wrong. He had the good sense to be completely wrong about the right issue, limits to resources, and as a result he is revered. What do errors matter? Why hang back? Activists occasionally claim that scientists are conservative in making statements because they are concerned that any mistaken claims will damage their reputations. A sideways glance at any of the mass of dire warnings and depressing forecasts, and a comparison of those forecasts with the evidence, shows that this is simply nonsense.

Instead, scientists are forced into extreme declarations in order to get any attention for their disciplines and this is not likely to change. Observers must make allowances for this over-sell in interpreting such declarations.

11
STORMS AND DISEASES

One 14th century storm, that is still remembered with horror to this day, is the Grote Mandrenke which is low Saxon for the great drowning of men. In January 1362 the storm swept across southern England, blowing down church towers, across the English Channel and into Holland where it obliterated whole parishes, (*Historic Storms of the North Sea, British Isles and Northwest Europe*, by H. H. Lamb and Knut Frydendahl, Cambridge University Press, 1991.) Writers of the time put the death toll at 100,000, but later analysis suggests that between 11,000 and 30,000 is closer. There are plenty more storms to choose from in the Little Ice Age. The great storm of 1703 is considered to be the worst storm to hit England, with winds in excess of 120 miles (192 kilometres) per hour wreaking enormous damage. Among other incidents it swept away the newly built Eddystone Lighthouse and its designer, Henry Winstanley, who happened to be visiting at the time, (*The Great Storm of 1703*, Ian Currie, BBC Online.)

Both of those events, you will note, are associated with the Little Ice Age and are just two of a long series of disastrous storms that swept in from the Atlantic during a time of unstable climate. Now we are being told that increasing temperatures and a time of stable climate will result in increased storminess. There will be more storms and they will be more severe. As in many other examples of lunacy detailed in this book it seems that global warming scientists want it both ways. What can we make of this?

To be fair, the storms cited above were only those that hit Europe and present-day scientists want to look at all storms in both the Northern and Southern Hemisphere. The North Atlantic Oscillation (NAO), an ongoing oscillation between a persistent atmospheric high pressure system over the Azores (South West of Portugal) and a persistent low pressure system over Iceland, governs storm tracks across the Atlantic. Different factors govern the hurricanes and typhoons and cyclones in different ocean basins. What are those factors and do they add up to an increase in general storminess? The answer to the last part of the question is yes and no, depending on which scientist you talk to. Scientific papers on this subject tend to contradict one another and, as one commentator recently noted, papers saying different things may even have the same authors.

To give you a flavour of the confusion in this area Greg Holland of the US National Centre for Atmospheric Research in Boulder and Peter Webster of the Georgia Institute of Technology in Atlanta produced a detailed statistical analysis of tropical cyclone activity that concluded there had been an increase in such activity and that the increase could be attributed to the recent increase in temperatures, ('Heightened Tropical Cyclone Activity in the North Atlantic: Natural Variability or Climate Trend?' Greg J. Holland and Peter J. Webster, *Philosophical Transactions of the Royal Society*, 2007.) Sim Aberson of the US National Oceanic and Atmospheric Administration, responded in the January 2009 Bulletin of the American Meteorological Society saying that the analysis was flawed. Using their approach, he said, the authors could have found a link between almost any two factors.

You will note that as everyone agrees that temperatures have gone up since 1850 - the real argument is about the cause of the increase – so messing around with hurricane statistics is not going to tell us much about what causes warming. It may tell us something about the consequences of further increases in temperatures, if scientists knew enough to forecast what might happen. But the area is in such confusion that it is better to leave scientists to work it out preferably after they have gathered a lot more data. While they are on the job, they may like to hammer out a coherent explanation of why decreasing temperatures in one era lead to more storms, and why increasing temperatures in another also lead to more storm.

For the moment the major storms are not paying much attention to forecasts of increasing frequency due to climate change. A recent NOAA press releases states that thanks to appearance of the El Niño climate cycle in the Pacific, the 2009 hurricane system was as slow as the 1997 season, ('Slow Atlantic Hurricane Season Comes to a Close', *NOAA*, November 2009.) The La Niña cycle which appeared recently may change that. A graph of tropical cyclone trends between 1970 and 2005 compiled by the Australian Bureau of Meteorology and on the BoM site does seem to not show any overall trend, at least nothing evident from just looking at it, except for a noticeable reduction in less severe storms. But then scientists can do wonders with statistics these days, so who knows what trend a sophisticated analysis may uncover.

One trend on which everyone can agree is that storms of all kinds have been causing considerably more damage in recent years in monetary terms, but that has nothing to do with global warming. The amount of coastal development has increased sharply in developed countries in recent decades, so there is now trillions of dollars worth of property spread all along the coasts of America and Australia, to take just two countries, where major ocean storms can get at it. Human efforts to limit the inevitable damage from these storms sometimes works but there have been spectacular failures, such as when Hurricane Katrina hit New Orleans in 2005. The city's levee system failed dramatically leading to widespread flooding which persisted for weeks as well as, of course, more than 1,300 deaths. The city has yet to recover, in part, as I understand it, because houses in the city are now very difficult to insure. The main problem with Katrina was that the city was unprepared and local building codes left something to be desired.

A better idea of the ongoing human-storm interaction can be gained from the results of Cyclone Larry which came ashore in northern Queensland in March 2006, causing widespread damage including shredding the local banana industry. Among other conclusions a subsequent survey of damage to houses in the Innisfail area says that more recent housing fared much better than older houses. Improvement in construction techniques meant that the roof was often better attached. The report also notes that many roller doors on garages in the area failed, with that loss often resulting in bits of the rest of the garage also blowing away. It concludes that aspects of the building codes need to be tightened, including construction details for houses on hill tops where the winds are likely to be stronger. ('Technical Report No. 51, Tropical Cyclone Larry Damage to buildings in the Innisfail area', School of Engineering, James Cook University, David Henderson and four others, September 2006.)

So again we have the problem of climate change lunacy distracting attention from the real problem, that of proper reviews of building codes and maintenance of storm and flood barriers. Messing with building codes and flood barriers, however, is not sexy. You cannot rally the troops and make impassioned speeches over those issues, but you can over climate change – if you ignore the changing science – so activists prefer that issue.

A similar effect can be seen in that other scare story de jour, that of tropical diseases increasing in severity and range because of changes in temperatures. Certainly there may be some increase in such diseases, if temperatures increase, but any change is likely to be swamped by the many other factors which also affect disease rates. Malaria is one case in point. Increasing temperatures are likely to increase the range and activity of the various species of mosquito which carry malaria, and the disease is undoubtedly a serious one in many, mainly tropical countries. A fact sheet from the US Centres for Disease Control and Prevention says that each year 350-600 million cases of malaria occur worldwide and more than one million people die, most of them from sub-Saharan Africa. The sad part is that many of these can be prevented right now and never mind climate change, just as the disease has been eliminated from both the US and England.

Although malaria is now associated with hot climes, it was prevalent in Western Europe until the late 19th century, and was only finally declared eradicated in America in 1951. The CDC material also says that the mosquitoes capable of transmitting the disease are still to be found in America, so if those mosquitoes ever get access to a few humans who have the disease another epidemic could result. The crucial point is that the mosquitoes are the disease vector, that is they spread it from person to person. The disease does not originate with them. They have to bite someone who has it and then later bite someone else.

Paul Reiter, of the Pasteur Institute in Paris, France, and acknowledged as one of the leading experts on the disease points out with some force that mosquitoes have been known to adapt to conditions in both the Sudan in Africa and Lapland in Finland, and transmit Malaria in both places, ('Global Warming and Malaria: knowing the horse before hitching the cart', *Malaria Journal*, 2008.) In England in Shakespearian times the disease was called the ague. He says that Denmark and parts of Sweden suffered devastating epidemics up until the 1860s. They were still occurring in some European countries and right across the old Soviet bloc countries from Poland to Eastern Siberia until well into the 20th century.

Reiter says that malaria died out naturally in many areas, despite the generally increasing temperatures, mostly due to changes in farming practices. These included draining of swampy ground but also the practice of keeping some animals such as pigs in large buildings at night, and keeping more of those animals. Mosquitoes like to breed in the farm building and so are more likely to bite the animals who do not get malaria. The use of the first anti-malarial drug quinine also helped, as did the discovery of the insecticide DDT, which has since been given a very bad reputation by environmentalists. None of this has anything to do with temperature, and Reiter has some harsh words to say about the use of computer models to simulate the spread of infection. He says that computer models are extremely useful in areas such as exploring the dynamics of transmission, but the interaction between vector (the mosquito) and host (people) is too complex for this approach. Another factor that may well be important in Africa is rainfall, as an increase in rainfall means more pools of water for the mosquitoes to breed.

Reiter explains: "Malaria vectors are, of course, still widespread in Europe and North America, and there are occasional cases of autochthonous transmission where an infected traveller infects local mosquitoes. Given the efficacy of anti-malarial therapy, however, and the increasing sophistication of disease surveillance, it is safe to say that there is no chance of significant transmission arising from such cases, and certainly no chance of the disease becoming endemic, at least under current economic circumstances." Reiter has proved a strong critic of the IPCC's disease projections, saying that his views and the views of other scientists with experience in tropical diseases have been disregarded in favour of computer models.

Instead of campaigning for unobtainable reductions in emissions, activists could make a much more effective contribution to world well-being in this area by pushing for funding for public health initiatives in those areas still stricken with malaria. But this is not as interesting and attractive as adding an increase in infectious diseases to the list of problems that will be caused by global warming, so they prefer the latter. A few activists have, in fact, being doing their best to block the use of an effective counter to the spread of malaria. They have been trying to persuade Africans not to use DDT, apparently not realising that the chemical was approved for use in limited circumstances by the 2004 Stockholm Convention, organised by the United Nations Environment Programme. The chemical had previously been banned, although the bulk of evidence against it existed only in the minds of activists. Now the chemical has been partially rehabilitated for use as a vector control, which means that users spray a little on bedroom walls to kill mosquitoes who come into bedrooms at night. This approach may have risks but it works a treat and avoids the problems of mass spraying which enables the mosquitoes to adapt. In other words, in any rational risk-reward analysis the use of DDT wins, but die-hard activists refuse to accept that there can be any situation in which DDT can be used to good effect.

Very similar comments can be made over the supposed increase in heat-related deaths to result from an increase in temperatures. Local factors, including the amount of money being spent on elderly care and the existence of very basic precautions such as issuing bulletins warning elderly people to stay in the shade and drink water on hot days, completely swamp any increase that may result from climate change. Once again we can see that concern over climate change, real or imagined, is distracting us all from the real issues.

12
SLIPPERY OIL NUMBERS

The concept of peak oil, unlike the other topics dealt with in this book, is not about climate but it has the same theme of industrial and economic activity being sinful. But instead of our consumption sinfully clogging up the atmosphere of our only planet, it is consuming all its key resources. Much the same groups of people who tell you that the earth is doomed by climate change will inform you, with utter conviction, that the end of the earth's oil reserves is in sight, and thrust sheaves of statistics and calculations at you to "prove" their point. Not only will it be very hot in the near future, it seems, we will feel much hotter as we will have to walk to most places.

Oil is in a different position to all the other commodities such as base metals (iron, zinc, etc), and coal in that it is obviously a key resource and its price history is different from the others. This is best explained by an illustration. In 1980, Paul Ehrlich, an American professor of population studies at Stanford University, made a famous bet with Julian Simon, a professor of business administration at the University of Maryland, that any five commodities named by Ehrlich would fall in value rather than increase over any period greater than a year. Ehrlich, who is best known for his 1963 book *The Population Bomb* which forecast global starvation starting in the mid-1970s due to population running ahead of food production, took the bet. He and some colleagues chose chromium, copper, nickel, tin, and tungsten over 10 years, and comprehensively lost. The resulting payment of just a few hundred dollars was hardly devastating but the incident illustrates the long-run price behaviour of almost all resources, including agricultural products. Prices tend to go down and producers remain profitable by increasing efficiency enough to stay ahead of the price curve.

A few remain unconvinced and some of those making these end-is-nigh forecasts, such as Ehrlich, have impressive credentials. In 1972 an informal group of like-minded businessmen, scientists and others who called themselves the Club of Rome produced *The Limits to Growth*. This book forecast a general breakdown in society in about the middle of what is now this century, in part because resources are going to run out. As we have seen commodity prices are not signalling an end any time soon (we will get to recent increases) and economists pay a lot of heed to price signals, but the book has its defenders who have since produced *Beyond the Limits* in 1992 and *The Limits to Growth: 30 Year Update* (Chelsea Green Publishing, 2004). According to those books we are on track to disaster, and the future will be like one of those dreadful apocalypse films occasionally produced by Hollywood where we degenerate into savages fighting over the few remaining tins of petrol.

This form of apocalypse has a perverse fascination for some people it seems, but there is little joy in the general run of commodities for these doom-criers. The mining industry and attendant analysts realised some decades ago that the amount of known, recoverable reserves of a particular economy such as coal and iron ore, has nothing to do with a foreseeable end to that commodity. Thus if there are, say, 80 years worth of iron ore reserves at present rates of production that does not mean the iron ore mines will stop operating in 80 years. The industry is always exploring for more. In fact, commodity prices are, within limits, better leading indicators of the size of proven reserves. The higher the price the more reserves eventually come to hand.

The rise of China in the past few years has upset the usual price-reserves-extraction process by sending prices for most commodities through the roof, with one result being a lot of investment in mining projects and a lot of exploration. This change in price regime tempted Ross Garnaut and his secretariat into looking at the issue of limits to resources in the previously mentioned final report of the *Garnaut Climate Change Review* in 2008. The report concludes, perhaps regretfully, that mineral reserves are not going to come under pressure until after "the environmental damage has already been done". The report also says, "any tendency towards exhaustion of reserves would raise prices, which would convert resources into reserves. It would also stimulate exploration, leading to expansion of reserves and the reserve base."

Food prices have also gone for a run of late, but this is generally attributed to an increase in demand – hundreds of millions more people can afford to eat – and the less likely cause of crop land being converted to the production of bio-fuels. In time this problem should correct itself.

So much for commodities but oil is generally considered to be different as its price has been rising since the 1990s. A little searching uncovers other points of concern on which a number of analysts of varying experience have pounced triumphantly. There you are, you can find trouble if only you look hard enough in the statistics. In fact, there are good reasons for thinking that oil is just like any other commodity except for being partially controlled by a cartel, the Organisation of Petroleum Exporting Countries. OPEC collectively controls 65 per cent of proven petroleum reserves, with Saudi Arabia controlling 35 per cent, at least as far as anything is known about reserves. Much of what seems odd about oil may well be due to OPEC, or oil producers in general, looking after their interests. Another part may be due to the oddities of oil itself. A glance at some of the recent debate shows that although apocalypse is unlikely OPEC's involvement means that oil prices may generally stay higher.

Those higher oil prices will mean more of those alarming price spikes that punctuate the price history of oil which, because it is such an important commodity, are noticed far more than price spikes in other commodities. A further complication is that politics intrudes into oil prices and supply far more than any other commodity.

One of those alarming price spikes famously occurred in 1973, thanks to the Arab members of OPEC imposing an oil embargo against the US and Western Europe, with that organisation also making intermittent efforts to control prices by imposing production quotas on its members. Prices touched $US40 a barrel at the time which works out at more than US$90 a barrel in 2009 prices. Another spike started in 2005 and culminated in prices pushing above $US140 a barrel in 2008, as the global financial crisis swept through the economic system. As prices were not much more than US$40 a barrel at the beginning of 2005, some concern was understandable. Also, because this spike happened at about the time some analysts forecast a peak in oil production, it was considered an indication that oil was about to be consigned to history, albeit after many decades when the last few drops had been wrung from dying fields.

The debate lost some of its urgency when prices collapsed again to around $US40 before recovering to somewhere a little south of $US80 at the time of writing, which is still quite high compared to the prices before the peak. Prices are still moving up after a (still largely unexplained) spike. Oil peakists, to use an unsatisfactory catch phrase for the people who push this concept, will also bring out graphs showing that world oil production has levelled off and that all the really big oil fields have been found and they are running dry. Our children will have to buy electric cars to drive their aged parents around.

Not so fast. Oil production has certainly levelled off in the past few years and that production plateau may have had something to do the unexpected spike in prices. It is also true that no new mega fields have been discovered for decades, but no one apart from the peakists are really sure what that means. As for the fields running dry, we will come to that.

In one form or another the debate about the amount of oil left in conventional reservoirs has been going on for many decades. Much of the hand wringing over the state of oil reserves and when oil production would peak – reach its maximum – dates from a paper by an American geoscientist M. King Hubbert written in 1963 and published as part of a study entitled *Resources and Man* put together by a committee in the US National Academy of Sciences and National Research Council. Hubbert's technique involves fitting a bell curve (the graph looks like a church bell – they are often seen in graphs of natural phenomena) to oil production. The oil peak was the top of the bell curve and production would decline thereafter. He forecast that oil production in the US would reach a peak about 1970, which was a reasonable call, although we will return to that forecast, and that oil production for the world would peak in the year 2000. He also estimated that natural gas production in the US would peak in 1980 and that coal production would peak well into the 22nd century or perhaps later, depending on a few unknowns at the time. I cannot recall anyone else trying to forecast peaks in gas or coal as opposed to oil, but in any case the last two forecasts are already wildly wrong.

In the 1990s, oil industry analyst Colin Campbell and petroleum engineer Jean Laherrère put together a more creditable forecast that oil production will peak in the first decade of the 21st century (about 2008) which was published in an article 'The End of Cheap Oil ' in the magazine *Scientific American*, in March 1998. The article made various points including the lack of new discoveries of the multi-billion barrel mega oil fields tapped by Saudi Arabia and other OPEC countries. Campbell and Laherrère gave reasons why they believed no more such large fields would be found, particularly in deep water. (The last part of that forecast has proved unfortunate.)

They also gave reasons why the recovery factor could not be increased substantially. In the 1960s, a rule of thumb for oil companies was that only 30 per cent of the oil in a field was recoverable, but by the 1990s that rule was 40 or 50 per cent. The two men say that trend is just a result of the way production is reported. As a field gets older more advanced technology is used to prolong its life, and petroleum engineers know far more about it and so can make more realistic estimates about the amount of oil in it. Either way, as readers will note, at least half the oil in any field is still left in the ground.

Campbell and Laherrère also used Hubbert's techniques, going to some trouble to work out available proven reserves. This is far harder than anyone outside the oil industry would suppose as the definitions for classifying reserves, such as the difference between "proven" and the less firm "probable" reserves, for example, varies from country to country. In any case, the reserve estimates published by the OPEC countries are mostly unreliable and show sudden, large jumps about the time member countries start increasing production. Under OPEC rules, at least for part of the time, the oil reserves held by a member country is a factor in setting its production quota.

Whatever readers may make of any of that, neither Hubbert nor Campbell and Laherrere ever said that the production peaks they forecast meant the end of oil as such. The more enthusiastic activists, the media and self-confident people in pubs, will sometimes tell you that crude oil production will grind to a halt very soon. They are confusing forecasts of a production peak with the end of production.

What the authoritative peak oil pundits say, and Campbell and Laherrère spell this out, is that they are forecasting a peak in production of easy lift, cheap oil. That peak then implies an eventual end, but there are still vast fields of what the oil industry terms "unconventional" oil, including the Orinoco belt of heavy oil sludge and tar sands and shale oil deposits in Canada, the former Soviet Union and Australia among other countries. The Orinoco reserves alone, the two men note, are estimated to contain a staggering 1.2 trillion barrels of oil. There are many billion barrels more in the Canadian sands. Hubbert's paper also mentions those unconventional deposits. The reason no one has bothered with them much, except in recent years as we shall see, is that easy lift oil is so much cheaper to extract and there is still plenty of it.

There are other, less satisfactory substitutes for oil. Cars can be converted to use Liquid Petroleum Gas comparatively easily, where there is enough of it available locally as there is in Australia. LPG is cheaper than ordinary petrol or diesel in Australia, but it is not as convenient as petrol as the tank has to be bigger and the distribution network is not as extensive. Does the local service station have LPG? Do service stations in country towns have it? But this is no problem for taxi fleets that do not stray very far from their home base. As a last ditch even coal can be converted into a form of liquid fuel for use in cars, which the Axis powers did during World War II, but it is expensive and the result is not as good as natural oil.

So whatever else may happen, and whatever faith we may place in the forecasts above, we will not be walking any time soon, or reverting to post-modern savages. The oil economy will continue for the foreseeable future. The immediate problem with oil production peaking and declining, at least as far as Campbell and Laherrère are concerned, is the extra expense of unconventional oil and the cost to the environment. In their paper they point out that the Orinoco sludge contains heavy metals and sulphur that must be removed and that sands and shales have to be strip mined. There are also the costs and time involved in ramping up production of those forms of oil while conventional oil production declines. They believed the switch from conventional to unconventional oil may lead to price jumps and disruption.

As oil is the most traded and analysed commodity (the next most traded is coffee), and future prices are a matter of interest, peak oil theories have attracted some interest over the years. Campbell and Laherrère are also in a different category to the wild-eyed greenies and the Club of Rome people who had previously been pushing peak oil, in that their analysis made some sense. They also had a point that no major fields had been found, just lots of smaller ones. So their thesis was kicked around for a while by academics and analysts. Then came the 2008 price spike with plainly puzzled analysts offering all sorts of explanations for the change, including peak oil. For activists, of course, peak oil was the only explanation possible and the end was nigh. We would pay for our excesses!

The debate bubbled along for some time, with one feature being a flood of popular books on oil. At the height of the price spike I counted six books on the oil crisis on the web site of a major book store chain, all by eminently qualified authors and all saying different things. These included *Oil* by Matthew Yeomans (The New Press, 2004), *A Thousand Barrels A Second* by Peter Tertzakian (McGraw Hill,

2006) and *Twilight in the Desert: The Coming Saudi Oil Shock and the World Economy* by John Simmons (John Wiley & Sons, 2005). The last book is worthy of note because according to it the Saudi oil fields, a mainstay of OPEC, were just about to run dry. Simmons presents a wealth of technical information, most of it impenetrable to the layman, to prove his point and it is often mentioned approvingly by peakists, but five years down the track the Saudi fields are still pumping away.

Analysts of all stripes added to the confusion by producing wild estimates of future prices. In 2008 Goldman Sachs analysts and oil industry figure Boone Pickens among others declared that oil prices would rise to $US200 with some forecasters pitching for $US300, just months before prices collapsed. Analysts who went in for peak oil placed the peak at anywhere from two years before the crisis, up to the 2060s. One group, the influential advisory company Cambridge Energy Research Association, put the peak in conventional oil production at 2040 but with unconventional oil adding quite a bit to that, and the peak extending over several decades. As is apparent from the material it produced at the time (CERA release, 14 November 2006) the company has no time whatever for the concept of peak oil saying that it can distort "critical policy and investment decision". You will note in all of this there is supposed to be a peak with some putting it soon and sharp but most pitching for later and as more of a long running plateau than a peak.

The debate arguably came to some sort of conclusion with the publication of an essay in the September 2009 issue of the American Economics Association's *Journal of Economic Perspectives* by James Smith, an oil and gas economist. In the essay, 'World Oil Market or Mayhem', Smith says a careful analysis suggests that the oil price peak was due to nothing more than good old fashioned supply and demand problems in an area where supply can only react slowly to changes in demand or, more crucially, a major interruption in one part of the world's supply. He does not specify those interruptions in the paper but there were plenty in early 2008, including saboteurs blowing up pipelines in Iraq and a strike by Nigerian oil workers. A few of those together may have been enough to trigger the spike. Then again, as he notes in the paper, it may be a classic bubble.

Whatever may have caused it the spike has gone away, but leaving prices generally higher. For that Smith blames both additional demand, from China in particular, but also on supply failing to keep pace. For the supply failure he lays a lot of the blame on OPEC. He says the organisation has failed to invest substantially in the production facilities necessary to keep pace with demand, and the reason for that failure is the organisation's need to maintain its dominating position in the market, including its ability to dictate prices. The prolonged higher prices of the 1970s and 1980s (once the embargo was lifted) was due to production restrictions by OPEC members but various members were always exceeding their quotas, often by up to 15 per cent.

If they don't invest in production then they do not have the means to produce the stuff and cannot cheat on their quotas. Controlling production is much easier. He also says that although the recent competition between analysts to pick an oil peak is "entertaining" it is essentially irrelevant to policy makers, as a peak – if and when it is reached – could have almost any results in the market. His arguments on that point are technical so we should move onto the issue of whether there is going to be a peak and, if so, when.

Smith also mentions Hubbert but is very uncomplimentary noting that his forecasts for coal and gas were wildly wrong, as was his forecast for the volume of oil to be recovered as opposed to when production might peak. He also points to extensive testing by others who found Hubbert to be of little use in forecasting petroleum extraction in individual petroleum basins. The two papers he refers to are *Testing Hubbert* by Stanford University academic Adam Brandt (*Energy Policy*, 2007); and papers by oil industry analyst Richard Nehring, 'How Hubbert Method Fails to Predict Oil Production in the Permian Basin' (*Oil and Gas Journal*, 2006). There are other oil and gas analysts, however, who swear by Hubbert.

Another, authoritative analysis of oil reserves by M. Adelman, a professor of economics at MIT and fellow academic the late Gordon Watkins, is also very uncomplimentary about efforts to forecast the end of oil reserves. The article 'Reserve Prices and Mineral Resource Theory' (*Energy Journal*, 2008), says that it is impossible to do so, with the main feature of the market being the role of OPEC in controlling prices by not doing very much about developing their resources.

They say, "Non-OPEC countries are competitors, who never have excess capacity, but produce all they can. Over time, they have kept increasing capacity and output, some tailing off as others grow. But OPEC, with far lower investment and operating costs, has actually reduced exports. Thus in the world industry for over 30 years water has kept flowing uphill: contraction in lower-cost areas, expansion in higher cost countries. Only a profit-maximising cartel of low-cost producers can explain this fact."

Although reliable statistics are hard to come by in the OPEC countries there is evidence, the two economists say, that around the Persian Gulf for many years all new reserves have been created in old fields. (Once a field is discovered it is common for geologists to add very substantially to estimates of oil in it.) Perhaps geologists in the area have been looking for new fields but been unable to find them? The trade journals do not indicate that a major search has been made, or any search at all, they say.

One hypothesis fits all these facts, including the recent plateau in oil production. OPEC countries judge oil in the known fields to be so plentiful that it's not worth looking for any more, and that by restricting investment in capacity they can keep prices high for what they have, the economists say. If the OPEC countries became competitors then all the countries would start investing in capacity to produce oil and prices would fall. They would be spending more money to get less. As it is, they are stringing out the life of their existing giant fields, adding wells every now and then when they identify new parts to it (oil fields are very complicated structures), but taking care to keep total production down.

"The price will cease to rise when OPEC find it unprofitable to keep raising, or when it becomes too difficult for OPEC to keep its output equal to the amount demanded at the current price. We offer no guess when this will happen. But the current price level has no relation to excess demand, nor to assumed resource inadequacy, of which there is no evidence."

Another piece of evidence in this debate comes from Richard Pike, head of the Royal Society of Chemistry. In an article in *Significance*, the educational journal of the Royal Statistical Society ('How much oil is really there? Making correct statistics bring reality to global planning') and in subsequent interviews he says that oil reserves have been massively understated due to a confusion over the way the reserves are added together. Oil reserves are usually quoted as the proven reserves, which are formally defined as the volume of oil engineers say can be extracted with 90 per cent assurance. Another way of stating this is that there is a one in 10 chance the field will yield less oil, and nine chances in ten that it will yield more. But if we are looking at a lot of fields together the P90 figure is more than just the sum of the individual reserves. Pike does a little more analysis to say that the usual figure for the world's total proven reserves of 1.2 trillion barrels (the total often cited by the peakists) can safely by doubled. In fact, the 2006 Cambridge Energy Research Association analysis cited above puts the total global resource base at 4.85 trillion barrels and likely to grow.

Those wedded to the concept of peak oil will not be convinced by any of the references above, and will instead point to 'Giant oil field decline rates and their influence on world oil production' by Mikael Höök of the Department of physics and astronomy at Uppsala University in Sweden and two others (*Energy Policy*, 6 June 2009). As we have seen a part of that production decline may be due to OPEC but the giant fields are undoubtedly going to decline, if they are not already in decline. If and when those giant fields pass away will that mean the beginning of the end, or just another phase in the oil industry? Perhaps not. One of the first uses of the early steam engines at the start of the industrial revolution was to pump up mining shafts that kept filling with water. All the surface deposits had been used so mining companies had to go below the water table to extract more of the ore, which was quite a problem in the late 18th century. No doubt someone told the mining companies at the time that this endless exploitation could not go on, and there must be a technical limit somewhere, particularly now that all the easy to access, surface deposits had been used.

As the big fields have become more developed oil drillers have been looking in other places for more oil, including further and further offshore. Adelman and Watkins say that before 1950 there was no drilling offshore or proved reserves. By 1975 they were drilling offshore wells in more than 300 metres of water, but analysts declared that to go any deeper was uneconomic because the oil rig would require too much steel. No sooner had they finished drawing that line in the sand than engineers went over it by devising methods of placing a well directly on the sea floor instead of building huge structures. Then geologists were able to use recently declassified US Navy data to see through what is called the "salt layer" in the Gulf of Mexico to discover oil-bearing structures underneath.

This has led to a string of recent discoveries. In September 2009 BP announced a field potentially as large as the largest of the North Sea fields with a declared three billion barrels of oil and analysts pushing for four billion barrels (*The Guardian*, 2 September 2009). The field was found in 1,250 metres of water after drilling 10.6 kilometres through the earth's crust. More announcements followed. In October Chevron, Devon Energy and Statoil, the Norwegian oil giant, declared that they had found 3 billion to 15 billion barrels in several fields 282 kilometres offshore, more than nine kilometres below the Gulf's surface (*New York Times*, 10 October).

Activists confronted with these reports tend to dismiss them as individually representing just a few months' supply, although that is a little harder for the Chevron discoveries. The same news reports estimate that if they live up to their potential, they will increase US reserves by 50 per cent. Another argument is that the oil is so deep it will cost a lot to extract. This is the line taken in *"Peak oil", the eventual end of the oil age*, a 2008 study produced by Jonah J. Ralston, a doctoral student in political science at Michigan State University at the time. He estimates that the cost of extracting oil from such deep wells can be three times the cost of conventional oil, and points to the environmental cost of extracting unconventional oil. Perhaps; but then the cost of taking mine shafts below the water table at the start of the industrial revolution was no doubt considered prohibitive. Mine owners of the time would have wrung their hands over the additional costs and wondered how they could make money out of it. Yet they did so and iron ore fell in price. As noted above the oil industry is different in that the OPEC fields could probably produce more at a lower cost, but until they do oil companies will have to develop these offshore fields and charge us more for oil. Oil prices will drift up but that is due to the OPEC cartel subverting the normal development of resources. Then there are the discoveries off the coast of Brazil, which are much larger than any of the others mentioned. Estimates of the so called pre-salt reservoirs range from 60 billion to more than 150 billion barrels. (*The Guardian*, 9 September 2009).

On top of all that, the technology to extract oil from existing fields keeps on improving, increasing the recoverable fraction mentioned above and defying Campbell and Laherrère's careful analysis. Then there is still unconventional oil. Because OPEC is keeping prices high parts of the unconventional reserves are worth the trouble, notably the Canadian oil sands with recent improvements in technology lowering extraction costs. There is perhaps 1.7 to 2.5 trillion barrels of oil in the Canadian province of Alberta, North of Calgary, of which 173 billion barrels can be recovered with present technology (*The Oil Drum*, 1 September 2009). But as of October 2009, prices were too low to permit new oil projects (Bloomberg, 8 October 2009). As a result, production to date has been relatively low. In 2008, according to the Canadian Association of Petroleum Producers, production from the province was only 1.2 million barrels a day.

From all of the above it is clear that anyone who was sufficiently worried about the concept of peak oil to wonder if there would be petrol for cars can stop worrying. The argument is still bubbling along in the mainstream but it is more about costs and about future supply, given that OPEC is not developing its resources. That was the main concern of Campbell and Laherrère, and in that sense they had something useful to say, although events have turned out quite differently from their forecasts. This often happens with forecasts. There are also the environmental problems associated with unconventional oil, but those concerns may be overcome with improved technology.

One of the key factors in the supply of oil in coming decades will be the time required for all the other sources mentioned to ramp up production in response to prices. As we have seen unconventional sources can produce a lot but investors will only put money into them when prices get past a certain point, and then it takes time for new production facilities to start working. It will also take time for the new Gulf of Mexico and Brazilian fields to start operating. More price spikes are possible. The International Energy Agency was recently reported as being concerned about further market mayhem, and also declared that many of the major fields were past their peak production (*The Guardian*, 3 August 2009.)

One related industry that has given up all talk of peak production, incidentally, is that of gas. A combination of 3-D seismic imaging and a new technology of fracturing of rocks known as hydro-fracturing to tap gas from shale deposits, previously thought too difficult and expensive to get at, is revolutionising the industry. One estimate is that the global reserves of gas have tripled in just three years (*The Daily Telegraph*, 11 October 2009).

As noted, there are plenty of peakists who will dismiss all of the above as quibbles, and declare that the world is teetering on the brink. They will also argue strenuously against the annual oil production forecasts issued by the International Energy Agency, drawing on any amount of reasoned analysis from this or that refereed paper to prove their point. In late 2009, there were also reports of a "whistle blower" in the IEA claiming that the US government has been pressuring the agency into using data from its own Energy Information Agency, with one result being a consistent over estimation of future production. This incident has been seized on triumphantly by the peakists. There you are, apocalypse is coming after all! But I am not sure that the incident is significant. I have deliberately avoided quoting heaps of statistics as even current production figures can be difficult to pin down, let alone a generally agreed figure for reserves or a useful estimate of future production. As we have seen, the market can be left to sort all this out, although it must be said there is a distinct risk of higher oil prices.

In any case, there is a handy rule of thumb we can adopt concerning peak oil that cuts through all that analysis. There are substantial deposits of shale oil around Gladstone and MacKay in central Queensland. According to the Australian Atlas of Mineral Resources, Mines & Processing Centres, for several years up to 2004 a pilot plant produced a total of 1.5 million barrels of oil from those deposits but has since stopped operating. There were complaints from environmental activists but also extraction costs were high. When that area starts producing again you will know that oil is indeed in short supply - either that or OPEC has been very successful at jacking up prices.

13
SHOW ME THE MONEY!

One reader's review of the useful book *A Primer on CO_2 and Climate*, second edition by American academic Howard C. Hayden, quoted elsewhere in this book, says "someone recommended this book to me. So I went here, and all I see are glowing reviews. Yet, if you check up on this retired professor, he sits on an organisation called CFACT that has received over $472,000 from ExxonMobil over that last seven years. CFACT has been critical of government regulation on many issues, including the o-zone layer, mercury emissions, global warming, toxic waste and the use of pesticides. While buying this $14.95 book helps supplement his income, it is pretty clear who is funding his retirement".

This kind of review comments in a reader review posted on Amazon are typical of the dirt flung by activists at anyone who dares to challenge their dearly held belief that the science on human induced global warming is rock solid. Also, like all such accusations, the amounts produced with a flourish by the global warming activists contradict the case they are trying to make, that big energy is bankrolling scepticism. The amount revealed works out to a little more than $US67,000 a year, which is trivial even in Australian terms let alone in America where CFACT operates, and never mind that it's been given to the organisation with which Hayden happens to be associated rather than directly to the scientist. The amount just looks large to activists.

The flip side of these accusations is that activists like to portray themselves and their scientific heroes as humble eco-warriors struggling against the odds and lack of funding to warn us all of impending doom. Again, a glance at any of the figures shows this to be complete nonsense. A paper recently released by the Science & Public Policy Institute in America estimates that in the 20 years up to end of fiscal 2009 the US Government spent more than $US79 billion on climate change research and technology. This included $US32 billion for climate research and another $US36 billion for development of climate related technologies. (*Climate Money*, Joanne Nova, SPPI July 2009.) Those figures are nominal, that is they have not been adjusted for inflation, but they are stupendous none the less, and the money is pouring in ever faster. According to one of the hundreds of emails hacked from the climate research unit at the University of East Anglia, CRU director Phil Jones was recipient (or co-recipient) of $US19 million worth of research grants over six years up to 2006 (a touch over $US3 million a year), a sixfold increase over what he had been awarded in the 1990s. ('Climategate: follow the Money', *The Wall Street Journal*, 2 December 2009).

SPPI is the enemy of climate change activism so they must be making that figure up, right? Again anyone who glances at the figures for themselves may consider the billions cited above to be an underestimate. When Kevin Rudd's Labor government won office in late 2007, one of its first acts was to set up the Australian Department of Climate Change. The Department's annual report for 2008-09 says that its income for the year was $82 million. That figure alone would be comparable to all the amounts paid out by energy companies to political organisations (with which sceptical scientists are sometimes associated) around the world for that year, and that is just the tip of the iceberg in Australia, let alone the other developed countries. Then there are the non-government organisations, that have embraced the cause of human-induced climate change, or have been created to fight it, such as Greenpeace, World Wildlife Fund, Friends of the Earth, the Environmental Defence Fund, Ozone Action, Clean Air Cool Planet, American for Equitable Climate Change Solutions and the

Alternative Energy Resources Association, to name but a few. There are more of these than you can shake a stick at, and all are adept at raising money to continue the fight against industrial emissions.

In a talk to the Political Economy Club in London in early 2010, David Henderson, a distinguished economist previously mentioned in the chapter on Model Mayhem quotes figures from the Copenhagen conference held in November 2009 that participants included nearly 23,000 representatives from NGOs ('The Climate Change Debate Today'). Of that number quoted a handful may have been nervous sceptics, but the overwhelming majority would have been activists. My guess at the number of active climate change sceptics, that is those who are prepared to do something such as create a blog or protest in front of parliament, is perhaps a few thousand. Those active at any time may be just a few hundred.

Then there is carbon trading. The SPPI paper estimates that carbon trading turned over $US126 billion (this is about the same as the 92 billion Euros figure I quote in the chapter on carbon). *The Wall Street Journal* article cited above also reports estimates that $US94 billion has been spent globally in 2009 on ethanol and other alternative energy schemes, and says that the latest European Commission appropriation for climate research was $US3 billion. These are awe-inspiring sums and the counting has barely started.

In contrast, in the same period surveyed by the SPPI in the report cited above Exxon-Mobil Corp has been repeatedly attacked for paying a grand total of $US23 million to organisations that are associated with sceptics. There is no figure or estimate for how much may have been paid directly to sceptics to undertake research but it would almost certainly be derisory. I can only recall one piece of research on climate issues declared to be funded by oil companies, and the report on it spent a lot of space pointing to that fact. In the present climate, pun intended, scientists would have great difficulty in accepting research dollars from any energy company. Further, if you scratch around on any energy company's internet site you will find some sort of statement about how the company accepts the need for action, and some green propaganda or other. They want to be seen to be green.

To counter-balance the vast number of NGOs pushing a climate change agenda, there are a handful of organisations formally structured to promote the cause of global warming scepticism which may (note the 'may') have a full time secretariat, as distinct from loose networks of sceptics in which someone may run a newsletter part time, or organise a blog. One is the SPPI mentioned above, another is the London-based Global Warming Policy Foundation, recently set up by former UK Chancellor Lord Lawson. Then there is the Non-government International Panel on Climate Change (NIPCC), which produces its own reports debunking the IPCC material. I have not used those reports but they are a handy guide to all that is wrong with the IPCC. NIPCC and the related Science & Environmental Policy Project (SEPP), founded by atmospheric physicist S. Fred Singer in 1990, have received some support from the conservative US-based The Heartland Institute. The National Centre for Policy Analysis in the US has also offered alternative views on climate change orthodoxy as has the Institute of Public Affairs in Australia.

Compared to the vast array of NGOs, government organisations, consultants who even earn their livings from the carbon orthodoxy and scientists looking for grants, this is a short list. But as far as activists are concerned it is far too long, and there must be energy money in there somewhere if only they can find it. In 2008 ExxonMobil reportedly gave $US125,000 collectively to The Heartland Institute and NCPA, and was continually attacked for its pains. A journalist who recently interviewed the GWPF's executive director, Benny Peisner, repeatedly asked about the size and sources of the foundation's funding.

Despite being unable to identify any funding that is not derisory or find more than a handful of organisations that may contest the global warming orthodoxy, activists continue to scream about how the debate is loaded against them, and even write books complaining that greenhouse science and national energy policies are really being dictated by the energy companies. For the rest of us it is clear that ravings about major energy company conspiracies to overturn accepted scientific truths are just that, ravings – another piece of lunacy in an area noted for its collective madness. The reality is that sceptics remain mostly unpaid part timers but, as we have seen, these dedicated volunteers have managed to score impressive victories against full-time climate scientists backed by massive resources. Having the truth on your side does help, sometimes.

As for the part about the scientists involved on the global warming side being humble and struggling against the odds let us all have a good laugh and get that one out of the way. Time and again, as we have seen, scientists and the IPCC have been caught out in gross errors which they have acknowledged only when they have had no choice, and while still calling those who caught them out harsh names. Both the hockey stick and tree ring incidents feature scientists refusing other scientists access to the basic data and source codes they have used for their work, although the scientific practice is now to permit access. Papers have been refused by journals purely because they contradict the scientific orthodoxy. That practice is common to all disciplines, but in climate change the exclusion is more systematic, and runs deeper. As shown in the so called "climategate" incident in which a host of emails from the Climate Research Unit at the University of Anglia were stolen and posted on the net, key global warming scientists have worked actively to prevent papers that contradict the orthodoxy from appearing in print. If they have appeared they will work to exclude them from the IPCC deliberations.

Faced with the distortions of science resulting from these efforts and the fevered conspiracy theory approach by activists to the debate, with both scientists and activists using the thinnest of pretexts to brush aside any sort of reasoned argument, this book will not make much headway. But it may help to persuade the more broad-minded of readers to at least wait a year or so to see what temperatures will actually do. That may be long enough to throw considerably more doubt on the issue and, as we have seen, make no difference at all to emissions either in the next few years or in the longer term.

As we have also seen temperatures have paused for the last decade or so and were even drifting down until the intervention of the El Niño effect which started in 2009. The UK Meteorology Office believes 2010 will be a warm year but, as discussed, those who watch the great oceanic climate cycles believe that temperatures will cool for the next few years. In particular, they say that this current El Niño should be much cooler, because the Pacific Decadal Oscillation is in its cooling phase. Very well, let us keep them to those forecasts. They are simple, straightforward and can be tested.

Then there is the sun, which has gone very quiet. Again as we have seen there are two sets of forecasts telling us that the sun should remain quiet for years to come. Perhaps the sun's activity is linked to the oceanic activity? This is one for the scientists to work out, but that lack of activity does not bode well for those who are looking for big temperature increases to prove their theories, or to raise panic over a looming hot and dry earth. For global warmers really need temperatures to go up. The computer models used to forecast climate are set up to amplify CO_2 increases, and as industrial gases have been pouring into the atmosphere temperatures should be going up. If they don't then, at the very least, we can draw the inference that climate is far less sensitive to CO_2 increases than some climate scientists have suggested. At best, and this is my preference, we may draw the conclusion that the whole carbon warming theory is nonsense from beginning to end.

Anyone familiar with the history of forecasting or prediction of all sorts or who knows anything about the limitations of computer models would be wary of the IPCC efforts in any case. The strange part about this extraordinary episode is that distinguished scientists of all types have rushed in to endorse these models although they, above all, should know better. Any of those scientists who had looked more carefully at the forecasts for CO_2 or at the extraordinary system of checking which the IPCC has passed off as peer review, or at recent temperature trends, or at any one of a number of other problems, should have headed for the hills. Everything has to go right for a forecast to be correct, and only one factor has to be wrong for the whole thing to come crashing down. But not only have scientists refused to acknowledge this, they have piled more computer models on top of the climate models to prove this or that disaster will result from projected temperature increases many decades into the future.

If temperatures continue their decline when La Niña passes then the vast empires of carbon which have sprung up in recent years will be gravely threatened, just as droughts during the Medieval Warming Period in South America helped pull down the Mayan civilisation. But it will take time. Carbon fantasies have taken such a strong hold over the public imagination and have proved so popular with activists that I do not expect to hear the end of this lunacy for many years to come, if at all.

During a recent debate on a Canadian public issue internet site, the Munk Debate, (www.munkdebates.com), leading sceptic Nigel (now Lord) Lawson, mentioned in earlier chapters, surprised participants with the fact that the earth had been cooling for the past few years, as it has. They declared it misinformation; if not completely wrong. In any case, look at how much warming had occurred before that, so what did the last few years matter? The participants had never been exposed to any of the sceptical arguments on global warming. Brief conversations with my colleagues about this book also indicate that they have not heard any of the recent problems with the IPCC. The last they heard were confident assertions that the argument has been settled. The better informed are vaguely aware that there was a period called the Little Ice Age.

Then there are many others who simply think we should not pollute so much, or use up fossil fuels, and science that suggests we are polluting too much is "good" science. One colleague I discussed this book with said that humans should be made to stop consuming so much because the earth had only limited resources. I suggested that such an assertion was really a matter of ideology.

"No, not a matter of ideology," she rejoined hotly, "we have to do it. Our resources are limited."

A lot of people feel the same way but we still live in democracies and voters need information on which to base decisions. This book is about giving them that information. Will voters forego income and consumption on the basis of science that is at best incomplete, and when their efforts will have no effect of any kind on the supposed problem? Or would they prefer to wait for better technology that may have much more effect on emissions? Or perhaps they may condone a switch to a combination of nuclear energy and gas, and never mind the activist horror over that approach with two technological wrongs making a right? I suspect a combination of the latter, if there ever is a case for cutting emissions at all. It is time to send the activists, and the activist-scientists, packing and start framing policies that may actually work.

INDEX

Aberson, Sim, 192
acid oceans, 30, 184
Adelman, M., 207, 210
agriculture, 10, 36, 118-120, 128
Akasofu, Syun-Ichi, 175, 176
An Inconvenient Truth - documentary, 37, 66, 98, 121, 168
Antarctic Climate & Ecosystems Co-operative Research Centre, 185
Antarctica, 34, 91, 175, 176, 177
Arctic Oscillation, 64
Arctic sea ice, 173, 174
Arctic, 45, 98, 168, 173, 174, 175
Argentina, 117, 130
Armstrong, J. Scott, 22, 30, 31
Atlantic Meridional Oscillation, 64, 99, 102
Australian Academy of Sciences, 30
Australian Department of Climate Change, 216
Australian Market Energy Operator, 152
Australian Productivity Commission, 122
Australian Treasury, 126, 128
Avery, Dennis T., 43, 52, 106
Beck, Ernst-Georg, 92
Bond, Gerard, 107
Booker, Christopher, 31, 50, 107
Briffa, Keith, 54
British Broadcasting Commission (BBC), 72, 191

calving, 177
Cambridge Energy Research Association, 206, 209
Campbell, Colin, 203, 204, 205
Canadian oil sands, 211
carbon dioxide hockey stick, 90
Carbon Dioxide Information Analysis Centre, 34, 73
carbon trading, 10, 18, 131, 137, 143, 217
Carlin, Alan, 78, 79
Carter, Bob, 28
Carteret Islands, 173
Castles, Ian, 87, 88
Cato Institute, 59
Center for Politiske Studier, 158
Certified Emission Reductions, 138
Clean Development Mechanism, 134, 138
Clean Energy Council of Australia, 153
climate determinism, 114
Climate Research Unit, 54, 57, 216, 219
climategate, 54, 216, 219
clouds, 7, 18, 62, 67, 77, 78, 80, 82, 107, 149, 162
Club of Rome, 200, 205
Copenhagen Conference, 89, 103, 132, 133, 134, 139, 143, 170, 173, 217
Copenhagen Diagnosis, 170
Coral reefs, 10, 182, 185, 188
cosmic rays, 107
Cyclone Larry, 193, 194

D'Aleo, Joseph, 60
dams, 134
Dansgaard-Oeschger, 43
Dawson, Graham, 129
DDT, 195, 196
Dessler, Andrew E., 110
Devils Hole, 40, 41
Diamond, Jared, 115, 116
Dubner, Stephen J., 16
E.ON Netz GmbH, 16
Easterbrook, Don, 103, 104
Eemian, 39, 40, 41
Ehrlich, Paul, 189, 199, 200
El Niño, 62,63,64,71, 72,103, 119, 184, 193, 219,220
Electricity grids, 153
Environmental Protection Agency, 78
Essential Services Commission of South Australia, 151
European Union Emissions Trading Scheme, 10, 131, 140
extinction rates, 180, 181
feedback effect, 77
Finland, 116, 117, 172, 195
Fort Collins Forum, 109
fossil fuels, 93, 94, 95, 148, 156, 163, 221
fossil leaf papers, 91, 92
Freakonomics, 16
Freudian psychiatry, 26, 27
Friends of the Earth, 139, 142, 216
Garnaut Climate Change Review, 127, 201
Garnaut, Ross, 125, 127, 141, 201
gas turbines, 149, 150
general circulation models (GCMs) 18
glaciers, 24, 45, 46, 168, 169, 170, 172, 173, 176
Global Historical Climate Network, 60
global temperatures, 5, 6, 9, 21, 40, 42,
45, 54, 56, 58, 59, 69, 71, 73, 99, 104, 105, 109, 132, 176,
Global Warming Policy Foundation, 116, 218
Goddard Institute of Space Sciences, 64, 68, 73, 101
Gore, Al, 37, 66, 121, 168
Grand Banks, 250, 187
Gray, William, 28, 79, 109
Great Barrier Reef, 128, 182, 184
Great Global Warming Swindle, 107
Green, Kesten, 22
Greenland, 47, 91, 113, 115, 168, 172, 177
Grote Mandrenke, 191
Hadley Centre, 57, 62, 71, 102
Hansen, James, 64, 68, 73, 79
Hayden, Howard C., 76, 215
Heartland Institute, 79, 218
heat island effect, 50
Henderson, David, 87, 88, 89, 194, 217
hockey stick graph, 47, 48, 49, 50, 52, 61, 70, 90, 108, 109
Holland, Greg, 192
Holocene, 5, 6, 39, 41, 42, 43, 45, 46, 51, 106, 108, 170
Holttinen, Hannele, 156
Hsu, Kenneth J., 106
Huang, Shaopeng, 48, 50
Hubbert, M. King, 202, 203, 204, 207
Hunt, Terry L., 116
Hurricane Katrina, 193
hurricanes, 98, 100, 120, 192, 193
hydroelectricity, 149, 153
ice ages, 5, 33, 34, 35, 36, 37, 38, 40, 41, 42, 97, 183
ice cores, 34
Idso, Craig D., 183
Institute of Antarctic and Southern Ocean Studies, 28, 79

Institute of Public Affairs, 218
intergalacials, 91, 108, 183
International Arctic Science Committee, 39, 175
International Energy Agency, 212
International Heat Flow Commission, 48
International Institute for Applied Systems Analysis, 40
Ioannidis, John, 25
isotopes, 34, 40, 52, 93, 105
Jaworowski, Zbigniew, 91
Jevrejeva, Svetlana, 172, 173
Jones, Phil, 54, 59, 216
Keenlyside, Noel, 99, 100
Kellow, Aynsley, 29, 50, 87, 88
Kininmonth, William, 28
Knappenberger, Chris, 69
Kukla, George, 31
Kyoto protocol, 131, 132, 133, 134, 135, 136, 137, 139, 143
La Niña, 63, 64, 103, 119, 184, 193, 221,
Laherrère, Jean, 203, 204, 205, 211
Lamb, H. H., 46, 76, 113, 114, 181
Latif, Mojib, 102
Laurentide Ice Sheet, 36, 42, 169, 170
Lawson, Nigel, 126
Levitt, Steven D., 16
Lindzen, Richard, 28
Little Ice Age, 6, 44, 45, 46, 47, 49, 50, 51, 53, 105, 106, 113, 114, 115, 117, 172, 174, 175
Livingston, William, 14, 15
Lockwood, Mike, 107, 108
Lorius, C., 76
Malaria, 194, 195, 196
Mann, Michael E., 47, 48, 49, 50, 52, 54
Matthews, Robert, 31
Mauna Loa, 84, 85, 86, 87, 92

McIntyre, Stephen, 48, 49, 50, 52, 54
McKitrick, Ross, 48, 49, 52, 54
McLean, John, 23, 29
Medieval Warming Period, 6, 44, 45, 51, 52, 53, 70, 106, 113, 114, 180, 221
Mendelsohn, Robert, 124
Methane concentrations, 7, 83, 84, 85, 86, 89
mid-Holocene maximum, 39 42, 45, 106
Milankovitch cycles, 37, 39, 40, 41, 45, 108
Millennium Bridge, 19
Monaco Declaration, 89 188
Moy, Andrew, 187
Munk Debate, 221
Murphy's Law, 24,81
Murray-Darling basin, 128
National Academy of Science, 78, 90, 102, 106
National Centre for Policy Analysis , 218
National Climate Centre, 28
National Oceanographic and Atmospheric Administration, 74
National Snow and Ice Data Centre, 173, 175
New Zealand Climate Science Coalition, 72
Nigeria, 116, 117, 130, 206
Non-government International Panel on Climate Change, 218
Nordhaus, William, 121, 122, 123, 124
North American Office of Atmospheric Administration (NOAA), 83
North Atlantic Oscillation (NAO), 102, 175, 192
North, Gerard, 49
Northwest Passage, 174, 175
Ollier, Cliff, 177

Organisation of Petroleum Exporting Countries (OPEC), 201
over-fishing, 187, 188
Pacific Decadal Oscillation (PDO), 64, 103, 119, 220
Paltridge, Garth W., 28, 76, 79, 81
peak oil, 199, 204, 205, 206, 209, 210, 211
Peer review, 8, 15, 22, 23, 24, 25, 26, 29, 50, 52, 81
Peiser, Benny, 2, 116
Penn, Matthew, 14, 22
Perry, Charles A., 106
Pielke Jr, Roger, 28
Pike, Richard, 208
Pilkey, Orrin H., 19, 21
Pilkey-Jarvis, Linda, 19, 20
Plimer, Ian, 29
Point Carbon, 137, 141
pre-historic climates, 98
Proudman Oceanographic Laboratory, 172
Purchasing Power Parity (PPP), 88
Rahmstorf, Stefan, 68
Rees, Martin, 188
Reiter, Paul, 195, 196
Royal Academy of Engineering, 156
Russian Institute of Economic Analysis, 59
salt layer, 210
Schnaars, Steven P., 32
Science & Public Policy Institute (SEPP), 23, 216
sea levels, 10, 21, 35, 42, 120, 129, 130, 167, 168, 169, 170-172, 177, 182, 183

Seasonal forecasts, 63, 71, 72
Segalstad, Tom V., 94, 95, 96
Shire of Hornsby, 167
Siddall, Mark, 169
Singer, S. Fred, 218, 243
Smith, James, 206
Solar and Heliospheric Observatory, 15
Solar concentrators, 161
solar cycle, 8, 13, 14, 17, 24, 106
solar magnetic activity, 15, 100, 105
solar magnetic theory, 41, 42
Southern Oscillation Index (SOI), 63, 64, 103
Spencer, Roy W., 28, 76, 77, 82
Stern, Nicholas, 66, 122, 125, 126, 127
Steve Fielding, 100
sunspots, 13, 14, 15, 105
Svensmark, Henrik, 107
Taylor, Peter, 82
temperature proxies 35, 47
The Day After Tomorrow - film, 42
The Garnaut Climate Change Review, 127, 201
The Taxpayers Alliance, 137
Thomas, Chris, 179
tidal gauge data, 171, 173
Topex/Poseidon satellite, 171
tree ring data incident, 48, 53, 219
Trenberth, Kevin, 54, 109, 110
trifluromethane or HFC-23, 141
tuneable parameters, 79
UK Energy Research Council, 155
UK Meteorological Office, 57, 71, 73, 102, 219

University of Alabama, 28, 60, 61, 74
University of Alaska, 175
US Academy of Sciences, 30, 164
US Centre for Disease Control, 194
US Department of Energy, 156
US National Research Council, 49
US National Solar Observatory, 14
Victor, David G., 140, 141, 142
Vostok ice core, 34, 36, 37, 56
Wara, Michael W., 140, 141, 142, 144
water vapour, 7, 18, 77, 78, 79, 80, 81
Watkins, Gordon, 207, 210
Watts Up With That, 60
Watts, Anthony, 60
Wave power, 161, 162
Weart, Spencer, 77
Wegman report, 49
Wikipedia, 50, 96, 106
Wilkins Ice Shelf, 174
Willis, Kathy, 180, 181, 182
wind farms, 10, 148, 150, 151, 152,
    153, 154, 155, 157, 159, 164, 165
wind forecasting systems, 155, 160
wind power, 138, 148, 152, 154, 156,
    157, 158, 159, 163
wind turbine, 152, 158, 159, 163
Wong, Penny, 100
Worm, Boris, 187
Younger Dryas, 42
Zenghelis, Dimitri, 125

www.ingramcontent.com/pod-product-compliance
Lightning Source LLC
Chambersburg PA
CBHW020639220526

45464CB00001B/219